D1153926

The
Things
We Love

The Things We Love

How Our Passions Connect Us and Make Us Who We Are

Aaron Ahuvia, PhD

Little, Brown Spark

New York   Boston   London

Little, Brown Spark
Hachette Book Group
1290 Avenue of the Americas, New York, NY 10104
littlebrownspark.com

First Edition: July 2022

Little, Brown Spark is an imprint of Little, Brown and Company, a division of Hachette Book Group, Inc. The Little, Brown Spark name and logo are trademarks of Hachette Book Group, Inc.

The publisher is not responsible for websites (or their content) that are not owned by the publisher.

The Hachette Speakers Bureau provides a wide range of authors for speaking events. To find out more, go to hachettespeakersbureau.com or call (866) 376-6591.

The author wishes to thank the following for the illustrations, photographs, or charts on pages 10–11 © 2014 Rajeev Batra, Aaron Ahuvia, and Richard Bagozzi; 42, copyright by Cristoph Bartneck; 45, courtesy of David Silberkleit, the Bugeyeguy, restorer and dealer of Austin-Healey Bugeye Sprites; 50, 52, 120, 171, 188, 233, 237, 244, 253, by Harshshikha Ambasta of Harshshikha Illustrations; 76, from Hur, J. D., Koo, M., & Hofmann, W., "When Temptations Come Alive: How Anthropomorphism Undermines Self-Control," *Journal of Consumer Research* (2015); 223, courtesy of Dawn Lowrey; 267, courtesy of Dani Clode Design, daniclode.com.

ISBN 9780316498227
LCCN 2021951323

Printing 1, 2022

To Ruth, Syd, Hannah, Aura, Isaac, and Jonah,
who taught me what I really needed to know about love

Contents

	Introduction	3
CHAPTER 1	A Many-Splendored Thing	7
CHAPTER 2	Honorary People	41
CHAPTER 3	What Does It Mean to Have a Relationship with a Thing?	65
CHAPTER 4	People Connectors	87
CHAPTER 5	You Are What You Love	109
CHAPTER 6	Finding Ourselves in the Things We Love	135
CHAPTER 7	Enjoyment and Flow	159
CHAPTER 8	What the Things We Love Say About Us	193
CHAPTER 9	Because Evolution	229
CHAPTER 10	The Future of the Things We Love	261
	Acknowledgments	279
	Selected Bibliography	281
	Notes	283
	Index	311

The
Things
We Love

Introduction

IN 1988, I HAD RECENTLY STARTED THE PHD PROGRAM IN MAR-
keting at Northwestern University's Kellogg School of Management and was fortunate enough to take a course with the
marketing legend Philip Kotler. (He is so well known that once,
when I was giving a lecture in Kazakhstan, an amazing three
hundred people showed up — not to hear me but to hear a lecture by a mere "student of the famous Phil Kotler"!) Professor
Kotler explained that marketing isn't just for businesses; it's for
everyone. Nonprofit organizations need marketing; politicians
need marketing; even single people looking for romance are
essentially marketing themselves, too.

I was in my twenties and single at the time. So although marketing was fairly interesting, dating was a lot more interesting.
And in the late 1980s, dating services were just starting to take
off. Professor Kotler agreed that I could write my term paper

on the similarities between marketing and dating. He told me about a professor of communications studies, Mara Adelman, who shared my interest. Together, she and I published a string of papers about the ways in which dating services were influencing romantic relationships. These papers attracted a lot of media attention, and I even ended up on *The Oprah Winfrey Show*.

That was great fun, but when it came time for me to pick a dissertation topic, I knew I needed to write something that would get me hired as a professor at a good business school. Studying dating services could land me on *Oprah*, but it wouldn't help me get a job. I had, however, invested years of work in becoming an expert on the psychology of love. Was there some way I could take advantage of all that knowledge?

Then it hit me. People talk about loving things all the time. Should we take this talk literally, or is it just another overwrought metaphor? And if people really do love things, what can the research into interpersonal love tell us about that? I was hardly the first person to notice that people love things. But to my good fortune, I was the first person to collect scientific data specifically on this kind of love, which in marketing circles came to be called "brand love." It has remained a professional interest of mine for more than thirty years.

Although I'm a marketing professor, my research has always been grounded in psychology, philosophy, and sociology. In my PhD program, there was a saying: "We study consumers the way marine biologists study fish, not the way fishermen study fish." In keeping with that idea, this book is written for anyone curious about what love is and how it works. You will find a science-based exploration of the psychology of loving things rather than a "how to" marketing book. That said, it's wonderful when businesses, artists, and nonprofit organizations focus on producing things that people truly love. The insights in this book will be useful to anyone committed to that sort of mission.

Even though the title of this book is *The Things We Love*, it's not really about things; it's about people. That's because, to a surprising extent, our love of things is really about creating our identities and connecting to the people we care about. In this book, you'll see how we use things to help us discover who we are, who we want to be, and how to become that best version of ourselves. We also use things to support our close relationships and to manage our reputations with the many people whom we may not be intimate with but who still matter to us.

This book provides scientifically grounded answers to common questions such as: How does our love of things compare to our love of people? Why do we love certain things and not others? Why doesn't everyone love the same things we do? Why do things play such a large role in our lives? What's the difference between loving something and just thinking it's really great? Does loving things detract from loving people?

Regarding terminology, I'm using the word *things* very broadly to mean "everything that isn't a person." Therefore, *things* can denote not only objects but also activities, as in "Do your own thing" and "Let's do something tonight." It's useful to talk about the love of both objects and activities because in practice they are hard to separate — e.g., your love of your phone is wrapped up in all the things you do with it.

The word *things* also denotes animals, for which I preemptively beg the forgiveness of my fellow animal lovers. I call animals things simply because I want to discuss our love for them in this book, and continually writing "the things and animals we love" is just too wordy.

I'd like to clarify one more phrase: "love object" is a psychological term that in principle means "anything a person loves," but in psychology it usually refers to a person (e.g., "The mother is the baby's first love object"). This may well be the first book in which the phrase "love object" refers primarily to things rather than people.

Whether you are a hobbyist, a nature lover, a marketer, a designer, an entrepreneur, a sports fan, or a music lover — or if you're passionate about something else — I hope you find something in this book that gives you insight into yourself and other people and helps you lead a richer life.

1

A Many-Splendored Thing

*It is good to love many things, for therein
lies strength, and whosoever loves much
performs much, and can accomplish much,
and what is done with love is well done.*

— Vincent van Gogh

THERE ARE MANY BOOKS ABOUT THE PSYCHOLOGY OF LOVE, BUT this one is different. This book is about our love for things, including the things we can't bear to part with and the things we love to do, each of which we have somehow managed to select from a staggering array of options. For example, on a typical shopping trip to Walmart, which stocks more than 140,000 items, we may walk past more products in an hour than most of our ancestors would have encountered in several lifetimes. And Walmart is small beer compared to Amazon, which sells more than two hundred million different things. Along with this enormous list of products to choose from, there are also the things we love that aren't for sale, such as our country and things we make ourselves.

We also can choose from a staggering list of things to do. These range from tried-and-true activities such as reading, listening to music, and gardening to surprisingly popular activities such as "gooming"* dogs to look like panda bears, making clothing out of duct tape, "noodling" (fishing with your bare hands), and competitive mooing. Among this vast array of objects and activities, we love a very select few.

In 2021, advertisers spent $755 billion[1] trying to get people to care about consumer products. Part of the reason they had to spend so much was that they were trying to overcome consumers' default relationship with things, which is one of mild interest and perhaps a little curiosity but no passion or strong concern. So as we'll see, when people do love things, something pretty unusual is going on.

What are the most commonly loved things? After more than thirty years of asking people what they love, I've noticed that some answers come up repeatedly. Loving nature or activities in nature (e.g., hiking in the mountains) is number one on this list. Love of nature is something just about all of us have in common, regardless of our other views and opinions: even right-leaning hunters and left-leaning backpackers can find common ground here. And nature, metaphorically, loves us back: there is a whole literature of scientific studies showing that being in nature, or even just looking at a potted plant, improves our happiness.[2] Loving God and loving pets are somewhat less common than loving nature, primarily because not everyone believes in

---

* Yes, that's really the word.

God or has a pet (although a fair number of people who don't have pets still love them). Our love for God and pets occupies something of a middle ground between loving people and loving things, a topic I'll return to later in the book. The other most commonly loved things include sports and the arts as well as one's home, car, cell phone, and clothing.

Yet these most frequently loved things only tell part of the story. There is a romantic notion that for every person on earth, there is someone out there who is the right loving partner for that person. Similarly, when it comes to the things people love, it never ceases to amaze me that for just about everything, there is someone who loves it. Collectors love all kinds of things — handcuffs, artistically decorated toilet seats, asphalt, cigar bands, funerary items, dental tools, condom tins, Kotex dispensers, and airsickness bags. These collectors number among their ranks Oscar-winning actor Tom Hanks, who spends his time away from the set hunting for vintage typewriters. Beyond collecting, people love lots of other unusual hobbies. Apple cofounder Steve Wozniak plays Segway polo in his spare time, while Google cofounder Sergey Brin lets off steam by flying around on a trapeze. Sometimes the things people love can be surprising because they seem so mundane. Did you know, for example, that Winston Churchill was also an accomplished bricklayer? He reportedly spent two years rebuilding and expanding his family's ancestral home with his own hands — a project that earned him an apprentice card in the Amalgamated Union of Building Trade Workers. And then there's Norwegian wood — not the song but the stuff people put in Norwegian fireplaces. In 2013, Norwegian television ran a twelve-hour program called *National Firewood Night*, which featured firelight aficionados discussing chopping and stacking techniques for firewood followed by eight hours of live footage of a burning fire — something that nearly 20 percent of all Norwegians tuned in to watch in its entirety.

THE LOVE OF THINGS QUIZ

What do *you* love? To find out if you love something, take the following Love of Things Quiz, painstakingly developed over several years by me and two of my frequent collaborators, Rajeev Batra and Rick Bagozzi, both in the marketing department at the University of Michigan's Ross School of Business. It has been published in top peer-reviewed scientific journals.[3]

To begin, choose an object or activity you love; anything other than a person is fine. Read the following thirteen statements with that thing in mind, then choose a number to indicate how much you agree or disagree with that statement: 1 means "strongly disagree," and 5 means "strongly agree." Then total your score.

LOVE OF THINGS QUIZ

		STRONGLY DISAGREE			STRONGLY AGREE	
1.	My overall feelings and evaluations toward this are extremely positive.	1	2	3	4	5
2.	I feel myself desiring it.	1	2	3	4	5
3.	I am willing to spend a lot of time, energy, or money on this.	1	2	3	4	5
4.	I have been involved with it in the past.	1	2	3	4	5
5.	I feel a sense of natural fit with this.	1	2	3	4	5
6.	I find this enjoyable.	1	2	3	4	5

7.	I feel emotionally connected to this.	1 2 3 4 5
8.	My involvement with this says something true and deep about who I am as a person.	1 2 3 4 5
9.	This helps make me the person I want to be.	1 2 3 4 5
10.	This does something that makes my life more meaningful.	1 2 3 4 5
11.	I frequently find myself thinking about this.	1 2 3 4 5
12.	This will be part of my life for a long time to come.	1 2 3 4 5
13.	If this were to go out of existence, I would feel a great sense of loss.	1 2 3 4 5

TOTAL UP YOUR SCORE HERE:

Scoring Guide

Where exactly to draw the line between the things we love and the things we don't is somewhat arbitrary. That said, I've created approximate cutoff points on the "love thermometer" (at right) to give you a sense of what your score means.

A score of 60 or above signals true love. A score between 50 and 59 is sort of love, a gray zone between love and not love. A score of 49 or below means not love.

Love thermometer scoring guide

65
True love
60
55 Sort of love
50
45
40 Not love
35

The Things
We Love

11

A REAL BUT DIFFERENT LOVE

To most people, it seems obvious that we can love things, but others disagree. For example, in 1988, Terence Shimp and Thomas Madden,[4] pioneers in the study of this topic, claimed that when people say they love something, they're just speaking metaphorically, so it isn't really love.

But we'll see plenty of evidence throughout this book that our love for things is really love. Still, that doesn't mean our love for things is the same as our love for people. There are many types of love: romantic love, platonic love, familial love, brotherly love, crushes, unrequited love, and so on. Notice that each phrase characterizes a particular type of relationship: for example, romantic relationships include sexual desire, whereas familial love does not. In each case, love is somewhat modified to fit the needs of the situation in question. Similarly, our love of things differs from other forms of love because it is partly shaped by our relationships to objects and activities.

In order to discuss these differences, I often take a comparative approach that contrasts our love for things with our love for people. I also discuss the psychological processes that create our love for things and their deep roots in our love for people. As a result, I'm going to talk a lot about our relationships with other humans. While my primary goal is to help you understand our love of things, if I can also help you understand your relationships with people, so much the better.

UNCONDITIONAL LOVE?

Although people do truly love things, we sometimes use the word *love* when we think something is excellent. For example, people say "I love your haircut" when all they really mean is "Nice haircut." When we use the word *love* this way, we're using a figure of speech called synecdoche, in which a word for part of something is used interchangeably with the word for the whole thing. For example, we say "Nice wheels" (the part) to mean "Nice car" (the whole), and we say "Get your butt over here" (the part) to mean "Get yourself over here" (the whole). Using the word *love* to mean "excellent" is a type of reverse synecdoche, in which love (the whole) refers to perceived excellence (one part of love).

The fact that we frequently use the word *love* this way is evidence that perceived excellence is an important aspect of love. That's why, for all love's mysteries, something predictable happens when people start talking about the things they love: they launch into a list of the love object's virtues — running is exhilarating and healthful; Teslas have amazing acceleration; and so on. In one study,[5] I found that this happened with 94.5 percent of the people I talked to, regardless of what the love object was. As one of them said about his favorite seafood dish, "If you die and go to heaven, that's what you get served."

Once people fall in love with something, they tend to exaggerate its virtues in the same way that parents exaggerate their children's talents. In this regard, our love of things is quite similar to the way we approach our romantic partnerships. One

1988 study[6] of dating behavior found that when a person falls in love with someone new, the largest single element of this new love is the belief that the other person is nothing short of magnificent. The author of this study, Bernard Murstein, called this aspect of romantic love "the Jack Armstrong factor" for men (named after a perfect-in-every-way hero of a 1930s radio drama) and "the Madonna factor" for women (named after the virgin Madonna, not the "like a virgin" Madonna). Of course, few of us are really that excellent. So as the emotional frenzy of new love fades with time, we maintain our love for our partners through a combination of accepting their faults and exaggerating their virtues.

Exaggerating the good qualities of the people we love has a big upside. The happiest married couples are not the ones who see each other the most accurately but rather the ones who maintain the greatest positive illusions about their spouses.[7] In fact, people in happy marriages tend to see their spouses more positively than the spouses see themselves. Similarly, exaggerating the virtues of the things in our lives makes us more satisfied with them.

Perhaps more surprisingly, positive illusions about something can not only lead us to think it's excellent but also increase our enjoyment of it. For example, if you offer people a taste of cheap wine and tell them it's really expensive, they'll usually say it's fantastic. It's easy to suspect that these people are lying about how much they enjoyed the wine. However, neuroscientists, using brain scans, have shown that the better people expect wine to be before they taste it, the more pleasure they experience when they drink it.[8] And this isn't just about wine. The same holds true for comic books: when people expect to like a comic book, they enjoy reading it more than they would have otherwise.[9]

The fact that people love things they see as excellent raises the issue of unconditional love. Once, after discussing this in a lecture, I was approached by a man who had a severely intellectually disabled brother. He said that by some measures, his brother was not "excellent," yet his brother was an "affectionate, gentle soul" and "a total sweetheart." My lecture attendee loved his brother very much. Does this mean that love isn't linked to excellence after all? Sort of.

First, it is interesting to note that as this man described his brother, he made a point of saying how affectionate his brother was. So partly, he *did* see certain aspects of his brother as excellent and listed those as reasons why he loved him. In addition, there probably was an element of unconditional love in this relationship, apart from perceived excellence. The noted love researcher Robert Sternberg calls this aspect of love a decision, meaning a decision to love some people, faults and all. This decision to love someone is closely associated with the religious concept of grace, which holds that God bestows love on us despite any lack of merit on our part. As the eighteenth-century essayist Joseph Joubert said, "Kindness consists . . . in loving people more than they deserve." I daresay that a successful life with other people, and especially a successful family life, would be impossible without a fair amount of this unconditional love going around. And as I argue throughout this book, loving something, by definition, includes caring about it in ways that go beyond "what it deserves," a feeling based strictly on the practical benefits that thing provides you.

This is, however, one of the common differences between loving people and loving things. We're much less forgiving of things than we are of people. When people come to believe that there is a major fault in something they love, they rarely continue to love it with the same intensity (although there may still be enough lingering warmth that they find it difficult to part with it). This isn't to say that we only love perfect things but that we put up with far fewer faults in things than we do in people. Our love for things is rarely unconditional.

There is one seeming exception to this, which, paradoxically, reinforces my main point. People sometimes love flawed things if those things are connected to persons they love, such as when a parent loves her child's drawing even if it's not exactly great art. But consider how much more parents love their own children's drawings than they do the drawings of their neighbor's children. The parents' love for the imperfect drawings, like the light of the moon, comprises reflected energy from another source — unconditional love for an imperfect child. And, that said, even though we may proudly post every one of our children's drawings on the refrigerator door, the ones we choose to keep in a scrapbook, the ones we really love, are the best ones. So even *that* love for a thing isn't without judgments about quality.

Further, people have a particularly low tolerance for imperfection in things that are purchased. For example, I interviewed someone[10] who loved his MP3 player when he first owned it but stopped loving it as soon as newer, better products came on the market. Even though the device hadn't broken and he was still happily using it, just knowing that something better existed was enough to reduce his love for it.

Does this mean that people only love products that are super-expensive top-of-the-line models? Thankfully for my budget, the answer is no. When deciding whether they love a product, people are willing to accept a few imperfections if they feel the product is a major bargain. But even for bargain-priced products, overall high quality is usually essential for something to be loved. Interestingly, when I've interviewed people about the luxury goods they love, the only complaint that comes up has to do with how much they cost. But this is quickly followed by a statement saying that the brand is worth every penny because it's so fantastic. And even here, any comment about price is really half complaint and half boast — a humblebrag about how wonderful the product must be to command such an exorbitant price.

While we have a deep-seated need to love people, loving things is more of a "nice to have" than a "need to have." This is probably because during most of human evolution, things didn't play a big role in our lives. As Nicholas Christakis, in his book *Blueprint*, explains about early humans, "If everyone in the group has no possessions, this does impose equality." Because we don't have a psychological need to love things the way we need to love people, we can be picky about things we choose to love; we don't need to settle.

The fact that consumers demand near perfection from the products they love has not been lost on businesses. The saying "They don't make 'em like they used to" is true — but not in the way it's generally meant. Since the 1980s there has been an enormous improvement in the typical quality of manufactured goods as companies strive to create more satisfying, or even loveworthy, products.

Falling in love with something often starts with our observation that it is excellent. But if we just think something is excellent yet don't feel a personal connection to it, that's not love. So while love may start with thinking something is great, it doesn't end there. Understanding our love of things is important to us, in part, because the things we love are important to us in ways that go well beyond their pragmatic usefulness.

LOVE IS DEEP

Is liking something simply a watered-down version of loving it? People occasionally use the word *like* that way, but for the most part the like/love distinction is much more interesting. For example, people like their romantic partners only slightly more than they like their friends, but people love their romantic partners a lot more than they love their friends.[11] So logically, loving can't simply be a stronger version of liking.

When I compared what people said about the things they loved with what they said about the things they merely liked, an interesting pattern emerged. People are around four times as likely to say that they truly love something if it connects to the deepest parts of who they are by helping them express their identity or making life more meaningful.[12] As anyone who has been in love will tell you, *love is a deep and profound experience*. Providing those meaningful experiences takes more than just being the best pencil in the box. Love requires a deeper connection. And if we consider again the most frequently loved things — nature, God, pets, sports, the arts, homes, cars, cell phones, and clothing — we see that these are all things that, for many people, contribute a sense of meaning or purpose to their lives.

I once interviewed a woman[13] (I'll call her Kathy) who loved "extreme couponing," which is basically Olympic-level bargain hunting. Extreme couponers find products that have several discounts available at the same time and combine them for even bigger discounts. These couponers then throw one discount after another on the product until they can, in one example from Kathy, "walk out of the store with $500 worth of products for $30." There is a sizable extreme-couponing community that connects through TV shows such as *Extreme Couponing* and through their own websites. Kathy, a professional accountant, loves this consumer sport the way Tom Brady loves football, and she plays to win with a similar determination. But there is an unexpected twist to her story. Many of the items that have the biggest available savings are personal-care products such as diapers, soap, makeup, and antiperspirant. When Kathy sees a particularly juicy opportunity, she buys the allowable maximum number of products, then takes them to drug rehabilitation centers, where she makes care packages for the women staying there. "A lot of these women are coming in off the street, and it means the world to them to get something like that." It's that last twist in the story, giving the products away, that imbues this otherwise purely consumerist activity with a deep sense of meaning. And it's that sense of meaningfulness, combined with her feelings of success and enjoyment, that leads Kathy to truly love it.

A similar example comes from "Sarah," who, at the start of a research interview,[14] listed her earmuffs as something she loved. I asked her to describe the earmuffs, and she started with the usual list of what she liked about them: "They are great and functional and they're fashionable and they're pretty classic so they haven't gone out of style." She added, "I wear them every day in the winter months, so I spend a lot of time with them." Despite this, though, after reflecting on her earmuffs a bit, Sarah changed her mind and said she didn't love them after all. Why not? "They're just earmuffs," she replied. They didn't connect to the things she felt made her life meaningful, such as "family, friends . . . maybe giving back to others."

I mentioned earlier in this chapter that a few researchers have claimed that people can't really love things. I sometimes hear this view from nonscientists who are curious about my work, too. In response, I ask if they think it is possible for people to love nature. In every case, they conclude that nature is an exception — that yes, people can love nature. After more discussion, they'll usually agree that religious people can love God and patriots can love their country. It's just the "little things," like cell phones, they don't think you can really love. Even though this view is incorrect, it reveals a real insight. Our love for people is a deeply meaningful experience. The reason these skeptics agree that people can love nature, God, and country is that they can see how loving these things might also be deeply meaningful. But to some people, it seems to sully and degrade love to say that you love your cell phone. Yet as we will see throughout this book, there are lots of ways that people imbue otherwise mundane objects, such as cell phones, with significance that goes well beyond their practical benefits.

LOVE IS SACRED

The author and TV personality Barbara De Angelis once said, "Love's greatest gift is its ability to make everything it touches sacred." This is echoed by the people I interviewed,[15] who say that the beauty and immensity of nature leave them feeling humbled and awestruck. For example:

To hear the quiet of a pine forest or to see mountains around me or between me and the sky — that makes me feel relaxed, calm, and very minute — like I'm not really here for very long compared to those mountains or that ocean.

These experiences aren't limited to nature. Medieval cathedrals were designed to engender a similar sense of awe, and architecture can still affect people in that way. But whereas nature often leads to a sense of humility, an awareness of one's own smallness beside the immensity of the world, architecture can provoke pride and confidence in humanity's creative power. As one man said[16] about buildings he loves:

Good architecture leaves me awestruck. It's my favorite form of art. It's beautiful, and it's functional. It impresses me that man can do something like that. That's the macho part, I guess. That man can tackle the earth and build a building 110 stories tall or something like that.

These contrasting experiences — humility and empowerment — remind me of one of my favorite philosophical quotations, from Rabbi Simcha Bunam of Pzhysha (1767–1827):

Everyone must have two pockets, so that he can reach into the one or the other, according to his needs. In his right pocket are to be the words: "For my sake was the world created," and in his left: "I am but dust and ashes."

The rabbi's point is that both perspectives are true, and each has its moments of usefulness.

A BRAND NEW RELIGION

Young people are increasingly moving away from traditional religion. In 2012, when the Pew Research Center asked people their religion on a survey and gave them choices such as Christian, Jew, Muslim, Hindu, and "none of the above," researchers found that the fastest-growing religious group in the United States was "none of the above," or the Nones, as they are sometimes called.[17] This trend was particularly strong among people under thirty. This doesn't mean that these Nones were all atheists, though. Many described themselves as "spiritual but not religious," and 68 percent of them said they still believed in God.

Spirituality outside of religion sometimes shows up in the things people love:[18]

> [Music is] spiritually stimulating in the sense of, you know, kind of bringing me out of this world into a realm that is kind of above my everyday day-to-day living.

> The ocean represents an embodied concept of God that I have. In our world there seem to be so few things that really have qualities of the divine, both sort of the largeness and the power. The ocean is such a big reminder of that, and I feel lovingly connected to that.

These people are clear about the spiritual aspects of their experiences, but not all religious attachments are that explicit. In keeping with the central role brands play in our culture, some people's love of their favorite brands includes religious aspects. For example, in my research I have found Apple to be far and away America's most loved brand. Russell Belk, formerly of the University of Utah, and Gülnur Tumbat, of San Francisco State University,[19] studied intense Apple devotees and found that the existence of their tight-knit community and their practice of proselytizing and converting nonbelievers resembled characteristics of a religion. Belk and Tumbat also found several quasireligious mythic belief systems among Apple devotees. These include a creation myth, in which the sacrifice of selling Steve Jobs's VW van and Steve Wozniak's HP calculator allowed them to create the first Mac in Jobs's parents' garage. There is also a satanic myth in which the forces of light (Apple) are locked in battle against the forces of darkness (at that time, IBM and Microsoft). And there is a myth of the resurrected savior, in which Steve Jobs (who came to Apple's first Halloween party dressed as Jesus) is forced to leave the company but then returns to restore the faith and bring salvation.

Apple fans who fell in love with the company in its early days, back when it was a tiny David battling big corporate Goliaths, see it as an outsider, an anti-corporate brand. As an old-school Apple user myself, if someone had asked me what non-tech products Apple owners were likely to have, I would have said Birkenstocks, granola, and pot. Boy, have times changed. Many young folks today see Apple as a luxury brand similar to Gucci and Chanel, with both the positive traits (quality and creativity) and the negative traits (shallow materialism) that are sometimes associated with luxury brands.

Whereas Apple products used to be owned by nerds and artsy types, in today's high schools iPhones are de rigueur in popular kids' cliques. Sadly, kids can even be bullied in school for owning non-Apple phones. This sort of obnoxious behavior by some of its users has cost Apple at least one high-profile sale. A *New York Times* story reported that Jerry Seinfeld's daughter grew upset when he bought her an iPhone, "calling it a 'mean-girl phone' and requesting something cheaper."[20]

If Apple is seen by young folks as a luxury brand akin to Gucci and Chanel, is that still compatible with quasireligious devotion? After all, mainstream religions usually value humility and chastity (even if religious leaders don't always possess those traits) — values that aren't exactly associated with most luxury brands. Research on brands and religion conducted by business school professors Ron Shachar, Tülin Erdem, Keisha Cutright, and Gavan Fitzsimons[21] sheds light on this question. They found that being into cool or prestigious brands can take the place of religion in some people's lives. Specifically, this study found that while all people have a need to express their identities, some do this partly through religious affiliation, whereas others seek secular forms of self-expression, such as using cool or prestigious brands. Consistent with this theory, the research conducted by Shachar and his colleagues found that when choosing what to buy, people who were religious tended to be less concerned with brand names than nonreligious people were. Interestingly, this was only true for product choices that might affect a person's public image (e.g., Ralph Lauren versus Target-brand sunglasses). But for product choices that would not affect a person's public image (e.g., Energizer versus CVS-brand batteries), there was no difference between religious and nonreligious people in how much they cared about brands. It seems that this tendency of nonreligious people to care about brands is really focused on meeting their need to express their identities, a need that, for other people, is partly met by religion.

To be clear, this research does not suggest that *all* nonreligious people are brand-conscious. In particular, people who are highly educated and politically progressive tend to be quite skeptical of both brands and religion, but this group is only a small part of the population. Across the whole population, it is still true that the less religious people are, the more they tend to use brands to express their identity. Moreover, many brand skeptics still use brands to express their identity — not so much through the brands they buy as through the brands they *don't* buy. For example, it's common for highly educated progressives to either reject clothing they otherwise like because it has a visible designer logo or in some cases to remove designer logos from clothing before wearing it.

A friend of mine, Rebekah Modrak, who is an artist and a professor at the University of Michigan's Penny W. Stamps School of Art & Design, did this one better. She printed some cloth badges that contained the image of a man playing polo (similar to the Ralph Lauren logo), but with the name Ralph Lifshitz (Mr. Lauren's original name) underneath. I sewed one onto a coat I had. Later, I was waiting in line at Macy's when I struck up a conversation with a woman next to me who turned out to be an executive at Ralph Lauren. I showed her the badge on my coat. She looked at it with stunned disbelief, then nearly lost it with laughter. Finally, she just said, "Oh, my God. Nobody calls him that."

A BIG-PICTURE OVERVIEW

What do people love? One way to answer that is with a list: nature, God, and so on. A better way is to say that people love things that have certain attributes. Rajeev Batra, Rick Bagozzi, and I conducted an extensive series of studies[22] that revealed thirteen components of the love of things, as indicated in the heart-shaped diagram on page 27. These correspond to the thirteen statements on the Love of Things Quiz on page 10. As the scoring for that quiz indicates, you don't need to score high on all thirteen questions in order to have your feelings classified as love.

Excellence. As I mentioned earlier, people think the things they love are excellent, even if they're seen through rose-colored glasses. This, in turn, helps us view them as valuable and important. In psychological terms, these positive evaluations are *cognitions*, meaning that they are thoughts rather than feelings. But love obviously includes a huge emotional component, not just thoughts. Which brings us to . . .

Positive Emotional Connection. In one study, Wendy Maxian, the chair of the communication department at Xavier University, and her colleagues[23] attached sensors to people that measured the response of their orbicularis oculi muscles (around the eyes), which automatically activate when a person feels like smiling. All the researchers had to do was show people brands

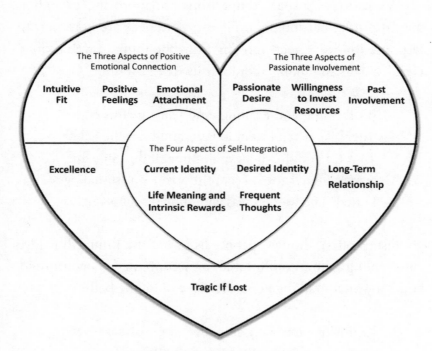

Aspects of the love of things

that they loved, and their orbicularis oculi muscles would begin to move, indicating *positive feelings*.

In my own research,[24] I have found that these emotional experiences come in two delicious flavors: exhilaration and relaxation. Here is a man talking about music's ability to produce exhilarating emotional experiences:

> [Music] gives me an emotional rush. It's intoxicating; you're just so into the song . . . both the lyrics and the music. At the end of it, it feels like I've just had sex or something, the rush is just so incredible. I think, goddamn, this really kicks ass.

While these sexual comparisons are more the exception than the rule, passionate, excited experiences are common in response to love objects ranging from shopping to waterskiing. Here's a woman talking about her love of traveling:[25]

> It's thrilling and exciting, and your dreams come true. Like when I went to Paris and I saw the Eiffel Tower, I couldn't believe it. After all this time and everything I'd read — there it was. I was so happy I cried. The friend I was with thought I was nuts.

Just as often, though, people associate the things they love with feeling calm, comfortable, and contented. For example, here's a woman talking about her love of taking baths:[26]

> It provides me with an inner sense of calm, because it is the one time where I totally allow myself not to become distracted or involved in things that require thought. It's just a time where I can be alone and relax. I use it to escape.

There can also be negative emotions associated with love. For example, jealousy can be part of romantic love between people. And although we are unlikely to feel a jealous anxiety that our new shoes might be cheating on us, other negative emotions can occur. The intensity of the love that fans feel for a sports team, say, is directly connected to the intensity of their pain when the team loses. In fact, the emotional devastation of big losses can be so extreme that if you add up all the joys and sorrows over the course of a season, some sports fans would surely be happier if they stopped following their team altogether. A few ex–sports fans I know stopped for this very reason. Yet many fans who have the bad luck of loving a losing team remain loyal not because it's fun but because the idea of abandoning their team feels morally despicable, like abandoning a loved one in their hour of need.

People often feel *an intuitive sense of fit* with the things they love. This sometimes creates a "love at first sight" experience — for example, when people who love their houses say that the house "just felt right":

> The first time I walked into this house, I knew we had to buy it. My husband felt very much the same way. It was sort of like unwrapping something, or unveiling something; each room sort of confirmed more and more this growing impression that this was the house we had to buy. So by the time we were in the back of the house, and the Realtor had left us alone, I just exploded with it: *"We have to live in this house — we have to."*

We also feel *emotional attachment* to the things and people we love, as if we are connected to them by an invisible emotional bungee cord that pulls us to them when we drift too far apart. There is a large body of scientific literature on attachment theory that addresses how these types of emotional attachments between parents and children affect children's relationships later in life. This research finds that children develop styles of relating to other people (attachment styles) based on how they are treated by their parents; these attachment styles continue to influence their social relationships as adults. For example, children who grow up in stable and loving homes develop a secure attachment style that allows them to have long-lasting and satisfying romantic relationships as adults. By contrast, children who grow up in emotionally dysfunctional families develop various types of dysfunctional attachment styles that interfere with their adult relationships.

It turns out that these attachment styles influence not only our relationships with people but also our relationships with things. Professor Vanitha Swaminathan of the University of Pittsburgh and her colleagues[27] found that not only were people with a secure attachment to their mothers more likely to form and maintain relationships with people as adults, they were also more likely to form and maintain relationships with brands. At the other extreme, children who cannot rely on their parents for love sometimes develop a fearful attachment style. As researchers Allison Johnson, Jodie Whelan, and Matthew Thomson[28] found, adults with fearful attachments to their parents can also develop "fatal attractions" to brands: they love the brand with great intensity early in the relationship, but if the brand disappoints them, they become enemies of the brand, obsessed with seeking revenge by writing vicious negative reviews for products they've never used, vandalizing company property, threatening to harm company employees, and other seriously nasty stuff.

When you combine the positive evaluations people have for the things they love with the emotional rewards they get from them and the emotional attachments they feel to them, it all adds up to a lot of . . .

Passionate Involvement. Passionate involvement includes a *passionate desire* to use the object or engage in the activity.

For example, one thing that loving people and loving things have in common is that both activate the region of the brain associated with hunger and desire. People are also *willing to invest resources* in the things they love. If it's an activity, we might spend time and money to take lessons. If it's an object, we might pamper it. Marketing professors John Lastovicka and Nancy Sirianni[29] found that people who loved their cars spent three times as much money caring for them as did people who merely liked their cars. And when people who loved their cars bought something car-related (e.g., new hubcaps), they often saw it as buying the car a gift. Finally, people are more likely to love something if they have a long history of *past involvement* with it. As one man said about his old car:[30]

> While that relationship may be coming to an end,
> I would say in terms of inanimate objects, I love
> that car. . . . It's been a traveling companion.

All these aspects of love — the perception that the thing you love is excellent, the positive emotional connection to it, and the passionate involvement with it — all combine to create . . .

Self-Integration of the Love Object. When we love someone or something, we expand our sense of identity so that the love object becomes part of who we are. The things we love become part of our *current identity* (who we think we are) as well as part of our *desired identity* (who we want to be). When I interview people, I often include what are called "projective questions," such as "If your stereo magically became a person, what kind of person would it be?" Because the things people love are part of their identities, they often imagine that their love objects would become a combination of who they are and who they want to be. For example, I interviewed[31] a freelance writer living on a limited budget who personified his Apple computer as a Victorian gentleman, "a person who matches my tastes in all things." But "he would probably be a little more into the creature comforts . . . better meals." He then went on to reflect, "I think I would have made a good Victorian gentleman. Yeah — someone who is hardworking and dedicated to what he does but wants money to be beneath him. Yeah, I really wish I was like that."

The process of integrating something into your identity doesn't happen instantly; your brain has to work at it. Our *frequent thoughts* about the things we love help integrate them into our sense of identity. Because we feel like the things we love are part of who we are, these things help give our lives *meaning*. The diagram on page 27 shows self-integration as a heart within a heart because it plays a particularly central role in love. Given that the things we love become part of our selves, it is not surprising that we also expect our connection to them to be a . . .

Long-Term Relationship. We've already seen that if our relationship with something has a long history and is currently active and passionate, it remains strong. Following this pattern, the stronger our love is for something, the longer we expect the relationship will last into the future. As the late psychologist Albert Ellis said, "The art of love is largely the art of persistence." But why not just love for today? Why should our future plans matter?

Love did not evolve so that we could bond with our cell phones. Love evolved in animals so that parents would take care of their offspring and, in some cases, take care of each other. One of the things that most surprised me when I was researching this book was how many animal species behave in ways that look a whole lot like human love. In many animals, offspring are on their own from the moment they are born. But in species whose young take a long time to mature, relationships have evolved that keep the parents motivated to feed and protect their offspring, just as human love does; in some species these relationships keep the parents together so they can both support their young.

Human children mature very slowly. It takes many years before they can survive without parental support, and their brain development isn't complete until around age twenty-five. Ideally, the love people feel for their mates and children should last a long time, although in real life this is not always the case. Similarly, love motivates long-term relationships with things as well as with people. But this motivation to stay together is weaker for loved things than for loved people, because we (quite reasonably) don't usually feel the same sense of moral obligation to things.

Given that we envision our relationships with the things we love as lasting far into the future, it makes sense that we consider them to be . . .

Tragic If Lost. When people try to decide whether they love something, they frequently ask themselves, "How terrible would it be if this were gone?" The more tragic the loss, the more they know they love it. This partly reflects the fact that the more important something is to us, the more we tend to love it. Although I've yet to measure this formally, I've noticed a pattern in which any given thing we love will often be important to us in a lot of different ways, such as being useful *and* fun *and* filled with sentimental value. As we'll see in chapter 5, the fact that people feel it would be tragic to lose something they love is also connected to the way these love objects become part of our sense of self, so losing them would feel like part of our selves was being torn away.

OBJECTIFICATION, HUMANIZATION, AND RELATIONSHIP WARMERS*

Loving things feels normal to us. But actually, it is deeply weird. To understand why, we need to start with the *social brain thesis*, which holds that because coordinating activity within families and tribes was so important to early human survival, our brains evolved to be more concerned with people than with things. As a result, the brain automatically sorts people and things into different categories and often thinks about them in different ways. In some situations, our brains even go so far as to use separate regions for thoughts about people and thoughts about things.[32] In particular, the brain normally thinks about things in a coolly instrumental way while thinking about people in a warm, emotional way, sometimes with love.

I call these different ways the brain normally thinks about people and things "default modes" of thought. The brain's default mode for thinking about things is very pragmatic.[33] Consider the vast number of things that pass before your eyes every

---

* Writing this book allowed me to step back from the individual studies I was working on and take a holistic look at the existing research about love and our relationships with things. Although this book relies on published scientific studies, it weaves these studies together into a new overarching theory, based on my and my colleagues' research, about how and why we love things. This section introduces some of that new thinking.

day that you completely ignore because they don't register in your brain as a potential threat (such as an oncoming car) or a nice opportunity (such as a delicious-looking cookie). Our default mode for thinking about people is to be more interested in them (our attention is drawn to people) and even to care about them a bit (if a stranger asks us for directions, our impulse is to try to help).

But the idea of a default mode also implies that thinking about people and things in these ways is not inevitable, since it is possible to override a default. To objectify people, for example, means to think about them as if they were objects. Sometimes this is harmless, such as when we are in a big crowd and our brain ignores most of the people in the same way it ignores most objects. But usually, objectifying people is a bad practice. The philosopher Martha Nussbaum[34] defines objectifying a person as seeing him or her as simply a tool for our own purposes. When we objectify people, we disrespect their autonomy, see them as lacking self-control, and even try to own them. I read a post on Reddit about a barista at a coffee shop during the COVID-19 pandemic who only had one customer. The barista asked the customer to put on a mask, to which the customer replied, "Why? There's no one here." On a conscious level, the customer knew that the barista was a person, but on an unconscious level, the customer's brain was treating the barista as just another part of the coffee-producing machinery.

In many ways, objectification is the opposite of love. In order to love things, we must reduce our objectification of them and think about them, at least partially, in ways that we typically reserve for people. In one unusual example, a woman I interviewed[35] talked about feeling hurt when the antique furniture she loved was *praised*. The furniture in question was a collection of family heirlooms and therefore was powerfully connected to her identity. She described herself as "hurt" and "offended" by some friends who had praised how nice-looking the furniture was. At first, I was confused by her response. But she explained to me that her friends didn't see the furniture as "more than just a pretty face": "These are things that I love. If other people don't realize that *they're not just pieces of furniture*, then that's what hurts my feelings." This is analogous to a woman being offended by a man who compliments her exclusively on the way she looks. It's not that she thinks being attractive is a bad thing. But she feels he doesn't appreciate who she is as a complete person and therefore feels dehumanized and objectified. The furniture lover felt the same way: despite the fact that the furniture was a collection of objects, she felt insulted when she thought they were being treated as such.

When we objectify people, we change the way our brains treat them. Princeton University researchers Lasana Harris and Susan Fiske[36] found that when we think about people for whom we have negative stereotypes, we activate the regions of the brain normally used for objects rather than the medial prefrontal cortex, which is normally used for thinking about people. Conversely, when we love objects, our brains treat them, at least partly, as if they were human. In a randomized, placebo-controlled study, the marketing professor Andreas Fürst and his colleagues[37] found that oxytocin, which plays a critical role in human bonding but is not normally involved in our relationships with things, helps create the bond between people and their favorite brands, just as it does between people in love. Similarly, Martin Reimann, of the University of Arizona, and

Raquel Castaño, of Tecnológico de Monterrey,[38] compared the regions of the brain people use to think about brands they love versus brands they feel neutral about. They found that a region of the brain called the insula, which is normally only used when thinking about people, was also being used when thinking about loved brands — but not when thinking about neutral brands. An analogous approach was taken by Reimann, Castaño, and research fellow Sandra Nuñez[39] in a study of how people experience pain. Previous research has established that thinking about people we love helps insulate us from physical pain. Reimann, Castaño, and Nuñez found that thinking about brands we love has a similar pain-reducing effect. More tellingly still, this pain-reducing effect can be increased by getting people to see a loved brand in human terms; conversely, it can be decreased by getting people to see the brand as just an object.

As I'll explain in detail in chapter 9, from an evolutionary perspective, it makes sense for people to care about objects exactly to the extent that the objects are pragmatically useful. If we are to be evolutionarily optimal, we should care about a useful object a lot, but if it stops being useful, we should immediately stop caring about it. After all, if our brains evolved with the "goal"* of having us love things, we would waste a lot of time and energy taking care of the things we love or even risk our safety to protect the things we love. But evolution wants us to focus on having lots of babies and taking care of *them*, not taking care of a lot of stuff.

---

* This is an anthropomorphic metaphor for evolution, which of course has no desires. The point is to stress the difference between our evolved love of certain people, which is an evolutionary advantage, and our love of things, which is an accidental byproduct of our ability to love people.

By contrast, one of the core defining features of love is that it involves caring about people and things even more than "they deserve." So loving things leads us to treat them in ways that aren't evolutionarily optimal for us.* That's why, from an evolutionary perspective, it makes sense that our brains are set up to prevent us from loving things by reserving love exclusively for people. The bottom line is that every time you love a thing, it involves a case of mistaken identity, in which the thing is being accidentally treated by your brain as if it were a person.

This creates a scientific puzzle that I call "the challenge of the social brain." Given that our brain is physically hardwired to form deep emotional connections to people rather than things, how do the things in our life ever become loved? Or, more concisely, *how do the things we love overcome the challenge of the social brain?*

There are three answers to this question. Which is to say, there are three situations in which the brain overrides its default way of thinking about things and starts thinking about them in ways it normally reserves for people. I call these three situations *relationship warmers* because they give emotional warmth to an otherwise cool, practical relationship between a person and a thing.

The first relationship warmer is anthropomorphism, which occurs when objects look, sound, or act like people. As a result, your brain starts thinking about these objects as if they were people. Anthropomorphism is addressed in chapters 2 and 3.

---

* That does not mean that loving things is bad. A lot of the best things in life aren't evolutionarily optimal.

The second type of relationship warmer is what I call "people connectors," which link us to other people. People connectors include photos of friends and family, gifts we've received, songs and objects that remind us of other people, devices such as cell phones that help us talk to other people, and lots of other things. When an object has a strong connection to another person, your brain starts to see it as part of this other person rather than simply as a thing. And because it is part of a person, it becomes eligible for love. People connectors are discussed in chapter 4.

The third relationship warmer is our sense of self. We have all embarked on a great life project to figure out who we want to be and how to become that person. The process of falling in love with something — whether it's a song that somehow catches your ear or a new house that just "feels right" when you step inside — helps us discern who we authentically are. Then, as we interact with the things we love, they become increasingly embedded into our identities, becoming part of ourselves. Because your brain normally loves your self, if things become part of your self, your brain stops thinking of them as mere things, and they become loved. Our sense of identity is discussed in chapters 5 through 8.

There is more to our love of things than just these three relationship warmers. But they are particularly important because they get your brain to think about a thing in ways it normally reserves for thinking about people. In the following chapter, I'll discuss the first of them: anthropomorphism.

2

Honorary People

*I caress the device in order
to induce it to do good work.*

— A COMMON RESPONSE TO A QUESTION
ABOUT GETTING TECHNOLOGY TO COOPERATE[1]

A CONTRIBUTOR ON REDDIT LAMENTED THE FACT THAT THEY had to get rid of their toilet after seventeen years of loyal service, and they "never had a chance to say goodbye." It's a little unusual for someone to want to say goodbye to a toilet, but I'll admit to waving and saying "Goodbye, house" as my family and I moved from our previous home to our current one. And if we shift from "leaving" objects to "loving" them, it is worth noting that in a survey conducted by Progressive Insurance,[2] 32 percent of the respondents said they named their cars and 12 percent said they would even buy their cars Valentine's Day gifts. This tendency to relate to objects as if they were people is called "anthropomorphic thinking," and it has a big impact on our love for things.

Anthropomorphic thinking means that at a conscious level, you know a thing isn't a person, but your brain responds to it as if it were. As I said in chapter 1, your brain normally reserves love for people. So in order to love a thing, your brain needs to treat it, at least partly, as if it were a person. Anthropomorphism is the most straightforward way of making that happen.

The iCat talking robot

Anthropomorphic thinking usually occurs when things "disguise themselves" as people by looking like a person, sounding like a person, or behaving like a person. Your brain turns out to be a sucker for these disguises, so they don't need to be particularly good to work. For example, the picture above shows a talking plastic cat. Christoph Bartneck, of the University of Canterbury, New Zealand, and his colleagues[3] conducted research in which participants interacted with this talking cat for a while and then received instructions from the researchers to turn it off. However, as soon as the plastic cat "heard" the researchers' instructions, it started pleading for its life, begging participants not to turn it off. As far as person disguises go, the toy's was pretty lousy. Yet when it begged not to be switched off, the participants' brains responded using many of the thought processes they would normally reserve for people, such as being reluctant to do something the cat didn't like and even reasoning with it in order to justify their actions.

In another example, Sara Kim, of the University of Hong Kong, and Ann L. McGill, of the University of Chicago,[4] created two images of a slot machine, similar to those shown below.[5] The only difference between the two slot machines is that the one on the left has lights on top that look vaguely like eyes and a mouth, and the handle is longer, so it looks a bit like an arm. The researchers measured the extent to which people saw themselves as capable of manipulating other people versus being vulnerable to manipulation by others. People who felt they were good at manipulating others preferred the human-looking slot machine on the left over the one on the right, whereas people who felt they were vulnerable to manipulation by others saw the slot machine on the left as being more likely to take their money.

This experiment shows that the tendency to see human-looking objects as having human qualities is so strong that even if an object is only a little bit anthropomorphic, the brain responds to it as if it were human. We can see this in studies that scan people's brains as they look at anthropomorphic objects. One study[6] focused on mirror neurons, which promote empathy by triggering within people the same emotions they see other people experiencing. The study found that these neurons fire not only when we look at people but also when we look at anthropomorphized objects. Another study[7] found that when people anthropomorphize objects, activity can be seen in the regions of the brain normally reserved for thinking about people — specifically, the medial prefrontal cortex and superior temporal sulcus.

When thinking about anthropomorphic objects, the brain uses some of the same clues to understand their "personalities" as it uses to understand people. Car companies, for example, consciously design the front ends of cars with various "faces." The evolutionary anthropologist Sonja Windhager and her colleagues[8] found that people assign human personality traits to cars based on what the front of the car — which designers sometimes call its face — looks like. The biggest difference between car faces was that some (such as the Subaru BRZ pictured on page 45) project a sense of dominance, arrogance, anger, and even hostility, whereas others (such as the Austin-Healey Sprite, also known as a Bugeye) seem kinder and even cute. Their research also showed that the primary attribute of the car that influenced these judgments was the extent to which it had a baby face rather than an adult face.

This Subaru BRZ has a "high-power" face.

The Austin-Healey "Bugeye" Sprite
has a friendly "low-power" face.

In a related study, marketing professor Jan Landwehr and his colleagues[9] found that when we look at human faces, we judge how friendly they are based on their mouths, whereas we judge how aggressive they are based on both their mouths and eyes. Similarly, people judge how "friendly" a car is based on the shape of its grille (i.e., its mouth) and how "aggressive" it is by looking at both its grille and its headlights (i.e., its eyes).

Why, you might wonder, would people want a car that looks angry and even hostile as opposed to happy and smiling? It would be reasonable, yet wrong, to assume that because people at the top of the social hierarchy are rich and powerful, they would smile a lot and rarely be angry or hostile. But it turns out that in group settings, the people in authority are the ones most likely to express disapproval, anger, and hostility, whereas the people at the bottom of the pecking order smile relentlessly, regardless of what they might be feeling inside. Therefore, we have come to associate angry, arrogant, and hostile facial expressions with "the boss," and some buyers, especially of luxury cars and sports cars, prefer vehicles that convey high social status and dominance. Some buyers also feel that these angry-looking cars are "strong" and therefore able to keep them safe.

A caveat is in order. Anthropomorphism encourages people to treat objects in a way that's similar to, but not identical to, the way they treat people. Your conscious mind knows that things aren't alive, and your conscious mind is hardly powerless. Furthermore, because anthropomorphized objects are human*like* but not human, our thinking about them tends to fall partway between the way we think about objects and the way we think about people. For example, researchers Maferima Touré-Tillery and Ann L. McGill[10] asked consumers to look at ads for a fictional company called CafeDirect, as illustrated on page 47.

In the first ad, the message comes from a smiling coffee mug, whereas in the second ad, the message comes from a smiling person. The researchers measured consumers' general tendency to trust other people in relation to how influenced they were by the ad. They found that consumers' general level

of trust in other people affected how much they trusted the smiling coffee mug (the more they trusted other people, the more they were influenced by the coffee mug). This shows that they were thinking about the mug as if it were a person. Not surprisingly, the researchers also found the same basic pattern for the ad showing the smiling person (i.e., the more consumers trusted other people, the more they were influenced by the photograph of the person). It's important to note, though, that consumers' general tendency to trust people had a bigger impact on their response to the ad showing the smiling person than the ad showing the smiling coffee mug. In other words, consumers' brains responded to the smiling mug as if it were somewhat human yet less human than an actual person.

Anthropomorphic thinking is a normal part of our everyday lives, even if we don't always recognize it. If you don't have an fMRI machine at home, one way to detect anthropomorphic thinking is to take notice whenever you talk to objects. Research[11] has shown that the objects people talk to most often are computers and cars, but in one study, people listed more than ninety other objects as being their conversation partners.

Another way to discover whether you've been engaging in anthropomorphic thinking is to notice whenever you feel angry at objects. Think of a time when a computer or other technological marvel wouldn't do what you wanted it to. Did you feel angry or merely frustrated? Anger and frustration are similar. The difference between them[12] is that we feel anger when someone has made choices that are causing us problems, whereas we feel frustration with things that are causing us problems but haven't *chosen* to do so: their actions are "just the way the world works." If a computer isn't working, and we feel not just frustrated with it but also angry at it, this means that at some unconscious level we are responding to the computer as if it were choosing to be uncooperative.

ANTHROPOMORPHISM LEADS TO LOVE

One way in which the brain treats humanlike things as human is that it forms emotional bonds with them.[13] When we bond with objects, our brains are thinking about them as social partners rather than simply as functional objects that provide practical benefits.[14] In keeping with this fact, our decision about repurchasing products that we imagine are alive is based more on how much we like their "personalities" than it is on their practical benefits.[15]

My research with my colleague Philipp Rauschnabel has shown, specifically, that anthropomorphic thinking is strongly linked to love.[16] In one study, we surveyed 1,100 German respondents about how much they loved their favorite fashion, chocolate, shoe, or shampoo brands. We also asked them about how high in quality they thought these brands were. Finally, we asked them about the extent to which they thought of these products as "having minds of their own" or being anthropomorphic in other ways. Using this data, we were able to see how strong the connection was between people's love for a product and whether they saw it as relatively high or low in quality as well as whether they tended to anthropomorphize it. Not surprisingly, the higher in quality people thought a product was, the more likely they were to love it. However, the connection between people's tendency to anthropomorphize a product and their tendency to love it was even stronger — in fact, it was a whopping *eighteen times as strong*.*

To cite a less scientific but much sweeter example, there was a story floating around the internet of a girl around five years old

---

* However, before anyone jumps to the incorrect conclusion that product quality doesn't matter to people, a clarification is in order. This research created a situation similar to what happens when people are deciding between their final two choices for a product. Consumers usually rule out the products they think are low in quality early in the shopping process. Once they are choosing between a few finalists, they usually think of all of them as pretty high in quality, so at that point they stop worrying so much about quality and focus on other aspects of the products instead. The final choice is often based on an intuitive feeling of attraction, which is part of love. And when people think about a product in anthropomorphic terms, they tend to love it more.

whose mother told her that their small potted houseplant, named Serena, wasn't getting enough sunlight. So the girl started taking Serena on walks around the block, holding the plant over her head to be sure it got enough sun. I strongly suspect that if the girl and her mother hadn't given their plant a name, it never would have occurred to the girl to treat it with such kindness. And, as the story pointed out, "kindness is love made visible."

When we think of an object anthropomorphically, it becomes an honorary person, which is a promotion of sorts. When this happens, a lot of the positive thoughts and feelings we have about people can rub off on the anthropomorphized object. That's one reason why in most situations, people see anthropomorphic products as better and higher in quality than similar nonanthropomorphic products.[17] We also see anthropomorphic products as emotionally warmer, meaning that they are more likely to be nice to us, which may be why we also tend to trust them more. Knowing this, companies design some self-driving cars, such as the Waymo Firefly (from Google), to look like people. Consumers see human-looking self-driving cars as more "mindful" and therefore more competent drivers.[18]

A drawing of the Waymo Firefly, Google's prototype for a self-driving car

While anthropomorphic thinking generally leads people to like things, under the wrong circumstances it can backfire. Researchers Sara Kim, Rocky Peng Chen, and Ke Zhang[19] conducted an experiment in which people played a video game and got hints for success from the game as they played. The research team found that consumers enjoyed the game less when the hints came from an anthropomorphized character and enjoyed it more when the hints simply appeared on the screen. The problem with having the anthropomorphic character provide the hints was that when players won, they felt like their victory was diminished because they received help rather than won on their own. When they received the same help but didn't see the source of the information as being alive, players felt their victory was fully theirs alone.

ANTHROPOMORPHISM AND MARKETING

None of this has been lost on marketers. Products such as Mr. Clean, whose brand is personified as an animated character, have been anthropomorphized in their marketing for more than fifty years. As authors Jing Wan and Pankaj Aggarwal[20] point out, over the years, "Mr. Clean has strengthened its human persona by appearing as the police officer 'Grimefighter,' as a 'changed man' when a new formula was introduced, and . . . in the 2013 'Origin' commercial, as someone who has been working hard from his childhood days to serve the cause of fighting grime to help others."

Today, companies are looking for new ways to get consumers to think about their products in anthropomorphic terms. Pankaj Aggarwal and Ann L. McGill[21] showed that if a company referred to a group of bottles as a "product family" and used a picture of bottles in various sizes (such as the ones below), consumers saw the bottles in human terms and liked the brand more than they did when they were shown a conventional product-line picture.

Another new way of getting consumers to see products as if they were people is to have the products speak to the users — as cell phones, appliances, and cars can do. Conversations with products often take place on social media, where companies tweet messages at consumers in the first-person voice of the brand or its mascot (for example, the Burger King himself might tweet rather than an employee of that company). Simon Hudson, of the University of South Carolina, and his colleagues[22] have found that these strategies may be effective because the more people anthropomorphize a brand they interact with online, the stronger and more intimate they feel their relationship is with that brand.

One of the most popular talking products is Siri, the voice of many Apple devices. It's so common for people to tell Siri they love her that Apple has a variety of preprogrammed replies, including my favorite: "I bet you say that to all the Apple products." All this talking to products reminds me of the guy who asked his cell phone, "Siri, I think I'm a nice enough guy, but I'm really striking out with the women. What should I do?" To which the cell phone replied, "My name is Alexa."

From a business perspective, marketers need to be careful, because anthropomorphism is a double-edged sword. For example, the more consumers see a brand as quasihuman, the more likely they are to see price increases as not just expensive but also unfair.[23] And consumers often get a lot angrier when anthropomorphized products fail than they do when conventional products fail.[24] This anger is partly attributable to the fact that when a product is seen as semihuman, it becomes morally culpable for its mistakes. If the product fails repeatedly, users feel as if the product is doing it on purpose.

These heightened feelings of anger and grievance can lead to a desire for revenge instead of simply a decision not to buy the offending product. Even before the existence of the internet, dissatisfied customers would, quite properly, complain about an underperforming company to their friends, thereby costing the company sales. But today, the internet has supercharged the ability of unhappy customers to make their voices heard. And the angrier customers are, the more effort they'll put into publicizing their complaints.

This anger can also get consumers to lash out directly at the offending product. In one study, 25 percent of PC owners admitted to physically attacking their machines. These assaults ranged from slapping them to throwing them out the window.[25]

WHY DO OUR BRAINS DO THIS?

Anthropomorphic thinking is a bit weird. So why do we think that way? Perhaps it is because much of the brain evolved at a time when human existence was even more of a group project than it is today. As a result, many parts of the brain evolved to serve as specialized tools for thinking about other people. You might have heard the saying that to a person with a hammer, everything looks like a nail. So to a brain that includes lots of tools for dealing with people, lots of things look like people.

That said, if a number of people all see a given object, some of them are more likely to anthropomorphize it than others. For example, you've probably noticed that children anthropomorphize objects a lot. Research[26] has confirmed this observation. What's more, our tendency to anthropomorphize things decreases somewhat throughout our lives until we get to the age of around sixty-five, when it starts to increase slightly. Interestingly, among adults, people who anthropomorphize things the most also tend to be highly imaginative, creative, and intuitive, all of which are traits that we associate with children.

There are two types of situations that are especially likely to lead to anthropomorphic thinking. First, our brains engage in anthropomorphic thinking in an attempt to ease feelings of loneliness. A famous example of this comes from the movie *Cast Away*, in which Tom Hanks's character, marooned alone on an island, creates an imaginary friend named Wilson out of a volleyball. In a similar real-life example, a man interviewed on the podcast *Hidden Brain*[27] talked about the effect that a severe sense of isolation had on him:

> It was a very lonely time. . . . I was in my apartment, and one of the fixtures in the apartment was a post. . . . The post became my friend. I would hug the post for all it was worth because I was getting some kind of feedback, physically. It was at that point I realized: I have got to do something. Because when you get to the place where you need to hug a post to feel something that you need, if that's not a wakeup call, nothing is.

While that is an extreme example, even normal, everyday feelings of loneliness increase our tendency to anthropomorphize objects and pets.[28] And in the aptly titled paper "Loneliness Makes the Heart Grow Fonder (of Robots)," researchers[29] showed that the lonelier people were, the more they liked a robot.

Second, our brains resort to anthropomorphic thinking when we want to solve a problem we're having with an object, such as figuring out how to get it to do something.[30] This is especially true when our first attempts with the object have failed. You may have noticed, for example, that if you turn the key in your car and it doesn't start, you just try again. But if you've tried several times and the car still won't start, that's when you start talking to it. Consistent with this common phenomenon, one study noted that when people wanted to get a computer to do something but didn't know how, 73 percent of them scolded it while 52 percent took a kinder approach by gently talking to it and encouraging it.[31]

All scientific researchers understand that pleading with, threatening, or showing affection toward a machine is not going to get it to "change its mind" and start working. However, there is divided opinion over whether anthropomorphizing a machine can, in a different way, help people solve problems. The claim that anthropomorphic thinking helps us do this is based on a fairly optimistic view of human nature. In this view, we wouldn't have evolved to do something if it didn't help us. So since people are more likely to use anthropomorphic thinking when they face a challenging problem, this means that anthropomorphic thinking is helpful when solving these problems.

The most plausible way that anthropomorphic thinking might help us solve hard problems is by engaging very powerful parts of the brain that specialize in thinking about people (a topic discussed in detail in chapter 9). This theory sees the human brain as a little like an airplane with economy and first-class seating. Your brain puts *things* in economy class, where a small number of mental resources are spread out over a lot of objects, but it puts *people* in the first-class section, where mental resources are much more plentiful. It could be that when we anthropomorphize an object, it gets the equivalent of a temporary upgrade to first class, where we can think about it with the powerful parts of our brains and come up with effective solutions. However, while this is plausible, we don't yet know if anthropomorphism actually helps people solve problems in this way.

I don't agree with the optimistic premise that because people resort to anthropomorphic thinking when they can't get machines to do what they want, this means that anthropomorphic thinking is usually helpful in these situations. Just because something is a common response to a problem doesn't mean it is a helpful response. For example, when we have disagreements with other people, we often see the problem as being entirely the other person's fault. We don't blame the other guy because it's a good way to solve the problem; we do it because it's emotionally easier than thinking about how we may also have contributed to the problem. Similarly, I don't think that anthropomorphic thinking is a good way to get machines to work. Machines are so different from people that anthropomorphic thinking just leads us astray. Why do we do it, then? Perhaps because most of us know a lot more about people than we do about cars, computers, and other objects. So when we lack the knowledge to solve a technical problem, out of desperation we fall back on the knowledge we do have and start treating the object as if it were human. Unfortunately, the "solutions" we come up with, such as reasoning with an object, are usually nothing more than distractions.

PETS

Pets aren't exactly things, but neither are they people. And we sure do love them. Mostly that's because they are friendly and fun, but it also turns out that they are good for our health. Even the National Institutes of Health, a conservative body on this type of issue, has concluded that interacting with pets is good for people, especially the elderly. This is because pets provide us with love, security, and a sense of purpose in life, all of which boost our physical and mental health. One study[32] showed that owning a pet increased the ability of people who have heart problems to stick with a rehabilitation program.

Not only can pets help you look after your health, they can also motivate you to achieve other goals and help you deal with stress. When you experience stress, your brain goes into fight-or-flight mode, which focuses all your attention on immediate short-term goals. This is good for dealing with a pressing emergency, but it means that all your long-term goals, from losing weight to finishing an important project, get ignored. Research shows that interacting with pets — or even just thinking about them[33] — can reduce your stress response and help you achieve your long-term objectives.

Unlike the things people love, animals do have thoughts and feelings. In this case, anthropomorphic thinking means seeing animals as more humanlike than they truly are. Studies[34] have found that people who are generally prone to anthropomorphizing things tend to have a strong personal connection to animals. Just as anthropomorphizing objects allows us to have a close relationship with them, anthropomorphizing animals can lead us to feel close to them. This effect was hinted at in a study[35] of animal mascots used by sports teams, which found that fans prefer anthropomorphized mascots that appear to be human-animal hybrids over mascots that remain true to an animal's actual appearance.

I'm about to discuss the ways people anthropomorphize dogs in a little more detail. But before doing so, I want to be clear that as someone whose family includes two wonderful little dogs — Noodle and Dumpling — I recognize myself in these descriptions of dog owners and pass no judgment on others. For example, I routinely talk to my dogs, saying things

such as "Dumpling, is Noodle stressing you out with his in-
cessant need to play?" This, as Gary Larson pointed out in
his classic *The Far Side* cartoon "What We Say to Dogs/What
They Hear," sounds to Dumpling like "Dumpling blah blah
blah blah blah blah."

In my research,[36] I've found that our tendency to anthropomorphize pets usually comes in two styles, each patterned after
a different type of interpersonal relationship. What kind of interpersonal relationship do you think each of these descriptions
is based on?

- **Person-dog relationship type 1.** Type 1 people prefer
  small, cute dogs, which they like to hold and cuddle; they
  like to buy their dogs clothing and toys; they believe their
  dogs should do as they're told and that it is the owners' job
  to mold and shape their dogs' characters. They see their
  dogs as innocent of the dangers of the outside world and
  hence vulnerable and in need of restrictive rules for the
  dogs' own protection. These people see their dogs as if the
  dogs were like human _____.

- **Person-dog relationship type 2.** Type 2 people like large
  dogs that have mature personae; they assume that their
  dogs are able to fend for themselves outside the home;
  they praise their dogs for being intelligent; they tend to see
  their dogs as their equals (or close to that); they believe that
  to maintain the right kind of relationship with a dog, they
  must respect the dog's wishes and not expect it to routinely
  do whatever they say. These people see their dogs as if the
  dogs were like human _____.

As you may have guessed, in the first type of person-dog relationship, dogs are seen as similar to children, whereas in the second type, they are seen as friends. In interviews conducted with dog owners by Australian researcher Michael Beverland and his colleagues,[37] people who saw their dogs as friends would sometimes let them off the leash, whereas people who saw their dogs as children were less likely to do so because they were concerned that their dogs might get "picked on" by other dogs or that they could get hurt because they didn't understand how to safely cross a street. One person who treated her dog as if it were her child made the dog a custom car seat with a specially made safety belt.

In these person-dog relationships, it's easy to see that some dog owners (often unconsciously) use their understanding of how parents relate to their children as a template for the way they relate to their dogs. This idea has received support from neuroscience. In 2015, researchers[38] conducted brain scans of women as they looked at pictures of their children, their dogs, other people's children, and other people's dogs. The results showed that the pattern of brain activity in women was very similar when they looked at their own children and their own dogs, but it was quite different when they looked at other people's children and other people's dogs. Apparently, when people refer to their dogs as their "fur babies," there really is something to that.

HOARDING: LOVE'S EVIL TWIN

For a long time, I resisted seeing a connection between loving things and hoarding them. But I now realize that although hoarding things isn't the same as loving things, hoarding and love do share some common ground. There are three main reasons why people hoard things, each of which has at least some connection to love.

First, hoarders have an irrational belief that the objects they hoard will be of crucial practical importance. Hoarders are genuinely convinced, for example, that someday, someone is going to show up and ask if they have a copy of the local newspaper from years before, and it will be terrible if the answer is no. So they see it as totally reasonable that they save every newspaper they've ever received in huge moldy stacks in their living rooms.

Hoarding is hard to overcome partly because despite these beliefs being baseless, they are tenacious. A rational person would point out that the odds of a hoarder, or anyone else, needing an old newspaper are extremely low, and if someone did need an old paper, it could always be found online or in a library. Unfortunately, one of the reasons that hoarding is classified as a psychological disorder is that these arguments bounce off hoarders' brains like a tennis ball hitting a backstop, and hoarders remain committed to their self-destructive beliefs.

This irrational and exaggerated belief in the potential usefulness of hoarded objects is similar to love but also quite different. Yes, people tend to hold exaggerated beliefs about the usefulness and importance of the things they love. But the difference in the extent of the exaggeration is enormous: in love, the exaggeration can be fairly large, but in hoarding, the exaggeration is often delusional and completely unmoored from reality ("Those used sandwich bags from twenty years ago are really going to be exactly what I need someday").

A second reason why people hoard things is that hoarders strongly integrate the objects they hoard into their sense of self,[39] which makes it feel like getting rid of the objects is like ripping out part of themselves. This is also something that hoarding and love have in common.

The third similarity between love and hoarding, and the one that I found most surprising, is that they both often involve anthropomorphism.[40] This is borne out by the fact that hoarders are much more likely than other people to agree with statements like "My possession can be thoughtful and sympathetic."[41] Why does anthropomorphic thinking sometimes lead to hoarding? Because when people see objects as human, they feel a moral obligation toward them. In one study,[42] children and teenagers saw an anthropomorphic robot that was on its way to being put in a closet and heard it say, "I'm scared of being in the closet." More than half these young people felt it was wrong to put the robot in the closet, showing that they felt a sense of moral obligation to it. Other research[43] has shown that the more people tend to anthropomorphize objects such as computers and motorcycles, the more likely they are to think it is morally wrong to harm them. One rental car company even found that when it placed human names in its rental cars, customers tended to treat them better.[44]

Part of this moral obligation stems from a concern for an object's "feelings." One study[45] found that people were more likely to turn off the lights when leaving a room if they had an anthropomorphic reminder from a human-looking lightbulb that said, "I am burning. Please turn me off," than they were if they had a conventional reminder. Among hoarders, this concern for the feelings of objects is much stronger than it is in most people, and it can lead them to buy things they don't need. As one hoarder explained:

> If I see one — one — packet of food left on the [store] shelf, I've got to have it, even though we don't need it or want it. Because I think that that thing will be really lonely left on the shelf, I hoard it.[46]

Once an object reaches a consumer's home, anthropomorphism also makes it harder for that person to dispose of it. Researchers[47] have found that when products are given anthropomorphic descriptions, consumers tend to keep them longer rather than buy new ones. This is good news for the planet; the environment would be a lot better off if we kept things longer and repaired them rather than continually replacing them. But hoarders take a good thing way too far, in part because they see things as alive. Marie Kondo, the famous tidying-up expert, advises people to say goodbye to their objects before getting rid of them. Before I understood the connection between hoarding and anthropomorphism, I thought that was just a quirky little ritual. But now I realize that it may be a crucial step in helping people part with objects they have anthropomorphized.

WHEN YOU LOVE things, it means you have a warm emotional connection to them, even though your brain normally limits these loving connections to people. In order for such attachments to form, the default cool, pragmatic relationship that you have with things needs to get warmed up. Anthropomorphism is our first relationship warmer. It works by disguising things as people.

When we anthropomorphize an object, it is easier to feel like we are in a relationship with it. This feeling is an important part of love. But it turns out that we can still feel a sense of being in a relationship with something even if it doesn't look or speak like a person. I'll explain how this works in the following chapter.

3

What Does It Mean to Have
a Relationship with a Thing?

I don't even like to use the word relationship.
I don't know what it means.

—RON SILVER (1946–2009)

RELATIONSHIP IS A WORD WE USE ALL THE TIME, BUT SOME OF us have a hard time defining it. Things get even more confusing when we talk about "having a relationship" with an object or activity. This chapter explains why you can't understand love without understanding what a relationship is, then explores what people actually mean when they talk about having a relationship with their hobbies, their shoes, and many of the other things in their lives.

IS LOVE AN EMOTION?

Almost everyone agrees that love is an emotion. The only people who seem to disagree are the scientists who study love. For example, as the all-star love research team of Helen Fisher, Arthur Aron, and Lucy Brown write, "Love is a goal-directed state that leads to a range of emotions, rather than a specific emotion."[1] To be clear, no one denies that love is an emotional experience. But most love researchers don't see it as a single specific emotion, like anger, joy, or fear.

There are two main arguments against the idea that love is an emotion. First, love lasts too long. Emotions normally last for minutes or hours. But love often lasts for years or even a lifetime.

Second, love is too emotionally complex to be *an* emotion. Instead, love involves a wide range of emotions. If you love a video-game console, the thought of losing it makes you sad; the thought of getting a new game fills you with excited anticipation; thinking about old games you used to play makes you nostalgic; and playing a game can make you feel, by turns, enthralled, disappointed, frustrated, and elated. In addition, researcher Sarah Broadbent[2] found that some people who love sports teams believe that love requires anger. As these die-hard sports fans saw it, if your team lost and you weren't angry at them for their bad play, you didn't really love your team. Since love involves all these various emotions, the argument goes, it can't be a single specific emotion.

Yet there are other prominent scholars[3] who argue that love *is* an emotion. And when they conduct studies in which they ask people how often they "feel love," no one has a hard time understanding the question. If there weren't a specific emotion corresponding to what we call love, wouldn't the study respondents have been confused?

Both views contain an element of truth but are incomplete. Like almost all English words, *love* has several different legitimate meanings. One is that love is an emotion, a strong form of affection that is associated with the release of oxytocin in the brain. Oxytocin is both a neurotransmitter and a hormone that is sometimes called the "cuddle chemical" because it is associated with sex, breastfeeding, trust, empathy, and close personal relationships. This strong affection, like all emotions, is fairly short-lived. It often occurs when you see a person do something particularly endearing, in which case you may experience what my wife calls a "love welling," bubbling up from your chest (perhaps explaining why people see love as coming from the heart), along with an inclination to tilt your head to the side and say "Awww," forming what scientists[4] call the "cuteness expression." Although the love emotion is often associated with cute kids and animals, we can also feel it for adults when they show us love or do something particularly admirable.

This affectionate love emotion is pretty rare in our relationships with things. Try this little experiment. Bring to mind a child you love and picture him or her doing something especially sweet. There is a good chance you'll experience a little love welling when you think of the child. Now bring to mind a thing you love, such as your car, your garden, dancing, music, or your favorite sports team, and pay attention to how it makes you feel. Thinking about a thing you love might make you smile, but you probably don't feel the same sense of love welling up in your chest that you did when you thought of the child. If you *did* feel love welling up, odds are that the thing you were thinking about was a cute animal, a stuffed toy, or a photograph of a person. These are exceptions that prove the rule. While animals, stuffed toys, and photographs aren't people, they are things that the brain treats a lot like people. So they reinforce the conclusion that the love emotion is tightly linked to people and doesn't translate well to things.

There is another legitimate meaning of the word *love* — the one love researchers are referring to when they say love isn't an emotion. Love is a type of relationship, similar to friendship and more or less the opposite of enmity. Like other relationships, the love relationship can last for decades, through better and worse, through thick and thin. And it involves more than a feeling. For example, love means doing nice things for the love object, thinking positive thoughts about the love object, and (often unconsciously) seeing the love object as part of who you are.

The fact that the love relationship and the love emotion are both called love suggests that there is a strong connection between them. I'm not aware of any research that has explored this question. But I suspect that feeling the love emotion toward something or someone makes you want to form a love relationship. That said, there isn't a perfect connection between the love relationship and the love emotion. First, it is possible to feel the love emotion toward things we don't love; it happens to me every time I see a kitten. Second, although this may sound like a contradiction, people frequently feel love for things (and people) without feeling the specific emotion of love. Usually when people say they feel love for a thing or a person, they are actually feeling happiness when something good happens to the love object, feeling empathy and sadness if something bad happens to it, missing it if it's not around, feeling excited about the thought of its return, and specifically in the case of sexual love relationships, feeling sexual desire.

When people talk about the things they love, they are almost always using the word *love* to mean this relationship rather than the love emotion. That's why when I use the word *love*, I, too, am usually referring to the love relationship unless I specifically say I'm talking about the love emotion. I should also note that love is just one of many relationships that people can have with things, a topic that has been authoritatively researched by Boston University's Susan Fournier.[5]

Because loving something means you have a particular type of relationship with it, we need to look at this relationship carefully.

ONE-WAY VERSUS TWO-WAY RELATIONSHIPS: MORE ALIKE THAN MOST SUPPOSE

Objectively, loving a person is normally a two-way relationship, whereas loving a thing is a one-way relationship. But subjectively, it doesn't always feel that way. In one study, Matthew Thomson and Allison Johnson,[6] of the business school at Queen's University, asked one-third of participants to describe a relationship with a romantic partner or friend, one-third to describe a relationship with a service provider (e.g., a doctor or hairdresser), and one-third to describe a relationship with a brand (e.g., Apple or Coke). Respondents then indicated the extent to which they agreed with statements such as "My relationship with X is mutual," which got at the question of whether the relationship felt one-way or two-way. The results showed that there was *no* significant difference between the people who were talking about person-to-person relationships and the people who were talking about person-to-thing relationships in terms of how two-way or one-way they felt their relationship was. This doesn't mean that there aren't any differences between loving people and loving things. But it does show that these two types of relationships have more in common than one would think. This rather odd finding has two main causes.

First, interpersonal love isn't always two-way. When people feel affection for other people, around 40 percent of the time they are merely thinking about other people rather than inter-acting with them.[7] Even though the people they are thinking of may in fact love them in return, at the moment they are thinking about their partners and feeling a surge of affection, their relationship with these other people is more one-way than two-way. Furthermore, research into couples' beliefs about their partners shows that each person's perceptions of the other tend to be so different that it's almost as if they are in two separate relationships.[8]

Second, our love for things can feel more two-way than it really is. This sense of having a two-way relationship with a love object is often the result of anthropomorphism. For example, Harvard Business School professor Youngme Moon[9] found that if a computer tells a user a little about itself (e.g., its processor speed), users will reciprocate by disclosing intimate information about their own lives. But even if a love object isn't partic-ularly anthropomorphic, people can still feel like they have a two-way relationship with it. Nonanthropomorphic things create a subjective sense that we are having a two-way relationship with them through their responsiveness and by providing emotional comfort.

Responsiveness

Responsiveness means that people pay attention to and respond quickly to one another. It has been shown to be important in creating strong marriages[10] and, in animal studies, even leads to attraction between rats.[11] When people love nonprofit organizations, or love companies that make their favorite products, responsiveness translates into the speed and care with which the organization responds to people's concerns. In my research,[12] I've found responsiveness to be important to both men and women. But an interesting study[13] of consumers' relationships with brands by marketing professor Alokparna Monga found that women care more than men do about how responsive a brand is.

People can love organizations partly because those organizations are responsive to them, meaning that the people who work at the organizations pay attention to them and their needs. Animals, too, can pay attention to us or ignore us (one great thing about dogs, for example, is that they are so happy to see us). But most of the things we love are objects or activities that can't pay attention to anything. What does responsiveness mean then?

Imagine a musician playing an upright bass. As the musician bows, the strings vibrate in response. Hearing the sounds from the strings, the musician adjusts his or her fingering and bowing, which in turn creates new tones. When this happens, the back-and-forth "dialogue" between the musician and the instrument creates a strong sense of relationship. Here is one bass player talking about his instrument:

> I have an upright bass. It's a beautiful piece of work. I can feel the vibrating strings. We've been very close for many years. . . . The experience of producing music is very personal and very intimate, and the bass is really a partner in that.

Emotional Comfort

Even though we consciously understand that objects don't feel love, we sometimes get an intuitive sense that the things we love return our affections. When people give us emotional comfort, this creates a sense of intimacy and can be an indication that they love us. Similarly, if objects give us emotional comfort, we can feel like they love us, giving us a sense that our relationship with them is two-way. For example, many children feel that their stuffed animals return their love because the toys give them a sense of comfort. Even when we've outgrown our teddy bears, we still find comfort in a wide range of things such as food, TV, music, exercise, alcohol, books, and nature, to name just a few.

THE SAFETY OF OBJECTS?

People can be judgmental, but that chocolate chip cookie is never going to criticize or reject you. I interviewed one person who, when jilted by a girlfriend, turned to his beloved computer for solace. His comments made clear that one benefit of the computer was that it would never leave him. Indeed, attachment researcher Jodie Whelan and her colleagues[14] have found that the more anxious people are about their interpersonal relationships, the more likely they are to form strong relationships with brands.

This is also a common aspect of our relationships with pets: one person I interviewed[15] said, "They don't question anything I do. They give me unconditional love. . . . It's very fulfilling."* But not everyone feels this way. I was struck some years ago by a Dear Abby letter in which "Larry in Delaware" was jealous of his cat's affection for his next-door neighbor. He complained that when his cat saw his neighbor, she would start purring, but the cat didn't do that for him. He concluded from this that his cat was being unfaithful. Abby replied that if the cat had fallen in love with his neighbor there wasn't much he could do, so he might as well accept the situation and move on.

And it's not just animals that may reject us. People who are lonely or isolated often make friends with hairstylists, bartenders, salespeople, and the like because they feel they won't reject them.[16] But even service providers, businesses, and products aren't always eager to return our affections. Marketers sometimes like their brands and products to "play hard to get" in the hope that this will make them more attractive. One common technique is to create the impression that a product is scarce. Hence the LIMIT 2 PER CUSTOMER signs next to some items,

---

* A friend of mine, Jeremy Wood, noted that this can also be true of our relationships with people who have passed away — as he *joked*, "My relationship with my mother has been great since she died."

suggesting that because they are offered at a low price, there won't be enough to go around. Marketers also like to create the impression that only the "best" people will get their products, and you might not measure up. Country clubs and condos often create this impression through a selective admissions process. Colleges bolster their prestige by boasting of their high rejection rates for applicants. This has led to the observation that colleges are like nightclubs: their attractiveness comes more from whom they keep out than from whom they let in.

Salespeople at high-end luxury goods stores are infamous for treating most customers in an aloof (or even rude) way, as if to say, "Owners of this brand are in an exclusive club, and you're probably not rich or famous enough to be a member." When I first learned how common this is, it struck me as not only obnoxious but also a terrible business practice that must be costing companies a fortune in lost sales. But Morgan Ward, of Southern Methodist University, and Darren Dahl, of the University of British Columbia,[17] found that when a brand is really expensive and customers see its purchase as "aspirational" (a big, special splurge), being treated in a standoffish way by sales staff just confirms that, indeed, the brand is a bit too good for them *and makes them want it even more.*

Ford took this to an extreme by making would-be buyers of its limited-production $500,000 Ford GT supercar fill out an application in order to own one. The company received 6,500 applications for only 1,350 cars. An application was most likely to be approved if the suitor was famous, had a lot of social media followers, was a well-known car aficionado, agreed to take the car to lots of car shows, and agreed to drive it regularly rather than save it in pristine condition as an investment.

Airline frequent-flier plans are probably the most common examples of products that give access to an elite club. If you fly often enough, you earn a privileged relationship with the airline that includes a special phone number, shorter wait times, entrance to a fancy lounge, better seats on the plane, and many other goodies. I'm not proud of it, but I'll admit it feels good to be an "elite" in one of these systems. I remember once having my suitcase crushed so badly by an airline that the metal frame was permanently bowed. I went to the airline's service desk, where the attendant told me that since one of the things that had broken was the handle, and the airline didn't guarantee handles, they wouldn't replace the suitcase. Then the attendant's phone rang, and she answered it. She said to another employee about the person on the phone, "It's Mr. So-and-So calling about his damaged bag." The other employee responded, "Oh, he's a *gold elite*; give him a new bag." My jaw dropped. But I quickly fished my frequent-flier card out of my wallet, showing that I, too, was a gold elite. "Why didn't you say so earlier?" the service person said. Then she immediately escorted me back behind the desk to a storage room that was stacked from floor to ceiling with brand new suitcases. "Take any one you want," she said.

When a salesperson treats us in an aloof manner, or when a car brand rejects us, these relationships can feel unsafe. Things can also feel unsafe when they are physically harmful to us, such as junk food that is tempting but unhealthy. On the one hand, people feel that these things are dependable sources of comfort. At the same time, they know that these things are harmful, leading some people to feel like they are in an abusive relationship.

Interestingly, research conducted by doctoral student Julia Hur and her colleagues[18] showed that anthropomorphizing sweets (say, by drawing a face on a cookie) makes dieters more likely to succumb to temptation. In Hur's study, all the participants identified themselves as being on a diet. Some participants were offered a tray of plain cookies with a note that just said, "This is Cookie," whereas other participants got a tray of cookies on which faces had been drawn, along with a note saying, "Hi, I am Mr. Cookie" (see the pictures below). Participants were told to eat as many cookies as they wished. The ones who got "Mr. Cookie" ate more cookies than those who got the ordinary cookie. Why? Dieters who got Mr. Cookie felt that "he" was encouraging them to eat more and was therefore partly responsible for their indulgence. In most cases this scapegoating of Mr. Cookie was unconscious. But to my surprise, there were several people who explicitly blamed Mr. Cookie for their choices.

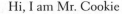

Hi, I am Mr. Cookie   This is Cookie

CLOSE TO YOU

I'll sometimes ask people I'm interviewing to rank the things they love and explain why they love one item more than another. All the following are typical answers. Which do you think is most common?

A. I love my phone more than my computer because my phone works perfectly but my computer sometimes freezes up.
B. I love my phone more than my computer because I use my phone all day and I only use my computer around three hours a day.
C. I love my phone more than my computer because my phone has a really cool-looking case and people compliment me on it all the time.

All these are important, but by far the most common answer is B, which equates the amount of love people have for something with the amount of time they spend using it. Weird, right? In one notable example, a woman said she used to love her iPod but doesn't anymore. Why? It isn't that the iPod changed or that she found something she loved more; it's just that she got a new job and replaced a long commute with a short walk. Now she uses her iPod less. She didn't stop using it because she stopped loving it; she stopped loving it because she stopped using it. What's going on?

I had an epiphany on this subject when I was reading about a questionnaire called the Relationship Closeness Inventory.[19] Researchers use the questions on it to measure how close two people are. For example, in order to determine how close a person taking the survey is with a friend, the Relationship Closeness Inventory asks the average amount of time, per day, that the respondent and the friend spend alone together. Things began to make sense: when people talk about how much time they spend with a love object, they're trying to get at how close they feel the relationship is. It brings to mind a definition of love offered by Elaine Hatfield and Richard Rapson of the University of Hawaii,[20] who wrote that love is "the affection and tenderness we feel for those with whom *our lives are deeply entwined*" (emphasis added).

This notion is useful when you're thinking about decluttering. If you have an object that you know you should get rid of, but you feel an emotional attachment to it, put it someplace out of sight for a year. When you stop interacting with it, your emotional connection to it will cool off, and it may become easier to give it away. When it comes to the things we love, absence doesn't make the heart grow fonder; it makes the heart grow indifferent.

MORAL OBLIGATION?

One of the primary attributes of relationships is that they stipulate the way people are supposed to behave toward each other. For example, in an employer-employee relationship, it would violate social norms for the employer to ask an hourly employee to do unpaid work for the company. But in a friendship, if two people aren't willing to do unpaid work for each other, they aren't really friends. Of all the various relationship types, loving a person has the highest level of expected obligation. If you're not willing to make serious sacrifices for someone you claim to love, it's not love.

Does this hold true for our love of things? When people love something, do they feel that they have an obligation toward it? And conversely, do people feel that the things they love should treat them particularly well because of the special nature of their relationship?

Altruism Versus Investment

Investment (giving up something now to get more later) is different from altruism (giving up something now to help someone else). People invest a lot of time, work, and money in the things they love. As one person I interviewed said about his beloved bass, "I sacrificed a lot to be able to play the instrument and to work with it."[21] But this type of sacrifice is an investment. People expect that their sacrifices will pay off in the future — when they enjoy skillfully playing an instrument, for example.

When we love people, we invest time and energy in our relationships with them and hope to recoup that investment as we continue to enjoy the relationships in the future. But we also make altruistic sacrifices for other people without a clear expectation that we will be paid back. And it's a good thing, too, because without daily altruistic caregiving from parents, children wouldn't make it to adulthood. But these types of altruistic sacrifices are less common in our relationships with the things we love. We usually don't feel the same moral obligation toward things as we do toward people, a view that I find perfectly reasonable. While it is unethical to treat a person as an object, it is usually fine to treat an object as an object.

Although we are less likely to make an altruistic sacrifice for a thing than we are for a person, it does happen. Think of environmentalists who protect natural areas by closing them off to development and tourism. They may lose the chance to visit these areas, but it's more important for them to ensure that the ecosystems remain healthy. Another interesting example, described by Stanford University's Aruna Ranganathan,[22] is that of artisans who give discounts to potential buyers in the hope that they will take good care of their handcrafted objects. The artisans sacrifice a little money to make sure their beloved creations get placed in a good home.

Brand Monogamy?

Our romantic love for people usually involves monogamy. Does this also apply to our love of brands? Some people — such as the man who posted a comment on the internet saying, "I am in the early stages of cheating on one of the longest-standing relationships of my (consumer) life. I have betrayed Apple" — seem to think it does. Others — such as the man who replied to the above post by saying, "The idea that you could even say 'cheating on Apple' is pathetic" — are of the view that humans don't owe objects or brands a promise of monogamy.

Pathetic or not, many people do see themselves as having monogamous relationships with their favorite brands. Cheating on a brand may mean one of two things. Some people think they would be cheating if they ever bought a competing brand. Others hold the view that being loyal to a loved brand means considering that brand first, but if it doesn't offer what they're looking for, they feel free to purchase something else.

Regarding cheating, I once heard a woman say about her husband's tendency to notice attractive women: "I don't care where he gets his appetite, so long as he eats at home." Businesses may be wise to adopt the same attitude toward their loyal customers. Marketing professor Irene Consiglio and her colleagues[23] found that, in many cases, when people consider themselves loyal to brand X but then "flirt" with brand Y by admiring or sampling it, even if brand Y impresses them, their attraction to brand X increases.

CARRYOVER EFFECTS

When behaviors that evolved for life with other people are applied to our relationships with things, or vice versa, I call this a "carryover effect." For example, researchers Oscar Ybarra, David Seungjae Lee, and Richard Gonzalez[24] found that when people feel insecure or unsupported in their romantic relationships, not only does this make them want to figuratively "shop around" for a better romantic partner, it also makes them want to literally shop around by considering a wide variety of product choices, even if the thing they are shopping for has nothing to do with their romantic relationships.

Professors Kristina Durante and Ashley Rae Arsena[25] have also shown that when women are in the most fertile part of their menstrual cycles, hormonal changes lead them to be slightly more attuned to attractive men in their vicinity. Again, this carries over into their relationships with things and leads women at that point to explore a variety of product options. However, when married women in the fertile part of their cycles are asked to take off and then put back on their wedding rings, this reminds them of feelings of commitment in their marital relationships. This, too, carries over to their shopping by making them less interested in exploring a lot of products.

These carryover effects influence not only how many products people consider but also what type of brands they find attractive. Consumers often see brands as having "personalities," such as being fun, cool, nerdy, caring, aggressive, mean, and so on. In terms of their personalities, brands tend to fall into two groups: exciting brands and sincere brands.[26] Consumers see brands like Hallmark as sincere, which means the brand is warm, caring, family-oriented, and often traditional. You know an ad is trying to give the brand a sincere personality if it features weddings, family gatherings, the national flag, people crying, nostalgia, cute kids, cute puppies, or cute anything, really. Consumers see other brands, like Porsche and the Bellagio resort and casino in Las Vegas, as exciting, which means the brand is fun, youthful, high-energy, and sometimes a bit transgressive. You know an ad is trying to give the brand an exciting personality if it features fast cars, dancing pop stars, scantily clad models, or people jumping off cliffs.

These sincere and exciting brand personalities are analogous to two types of people in the dating world. Exciting brands that are attractive but not always dependable are like sexy "hotties," or, as Boston University's Susan Fournier calls them, flings.[27] Sincere brands, on the other hand, are like "keepers" — the nice guys and gals who may not make your heart pound fast but offer many long-term rewards. Fournier and her colleagues Jennifer Aaker and S. Adam Brasel[28] found that consumers' relationships with exciting brands tend to look like flings—i.e., they develop quickly but then falter. By contrast, consumers' relationships with sincere brands, like our romantic relationships with keepers, tend to deepen over time.

Here, however, is where it gets a little bit weird(er). It has been said that "the quarrels of lovers are the renewal of love."* Could this also carry over to the things we love? It turns out that if consumers have a bad experience with a brand (e.g., a product breaks or the customer service is poor), this harms their

---

* This bon mot has been attributed to the seventeenth-century French playwright Jean Racine as well as to Terence (second-century BCE playwright, known for his comedies).

relationships with sincere brands, but like a lovers' quarrel, it can reinvigorate their relationships with exciting brands. This seems to be attributable to a belief on consumers' part that with a sincere/keeper brand, the problem should never have happened in the first place. By contrast, with exciting/fling brands, having the problem wasn't that surprising, but fixing it was. In these cases, people felt closer to the exciting/fling brand after it fixed a problem than they would have if the problem had never happened.

The ways people relate to exciting and sincere brands also differ based on how secure they feel about their own attractiveness. When dating, people who feel anxious and worry about being rejected often want to date people they see as being accepting. Consistent with this, attachment researcher Vanitha Swaminathan and her colleagues[29] found that people who are anxious in their dating relationships tend to favor sincere brands — the nice guys and gals of the brand world. By contrast, some people who are not overly worried about rejection are uncomfortable with commitment and intimacy. They tend to favor sexy "bad" boys and girls, who aren't likely to demand much commitment from them. These dating preferences also carry over to their love of things: they tend to favor exciting brands, the bad boys and girls of the brand world.

For readers interested in the social sciences, I would note that carryover effects take research in heuristics in an interesting direction. Heuristics are mental shortcuts people use when making decisions. They lead us to make decisions that are in most cases reasonably close to the right ones but in some cases significantly off the mark. Carryover effects are similar to heuristics in that they can lead people to do irrational things.

But carryover effects and heuristics exist for very different reasons. Heuristics, which provide ways of reducing the brain's workload, evolved because our brains have a limited capacity.[30] Carryover effects have a different backstory. Instead of developing as laborsaving devices for the brain, they emerged because the human brain evolved mostly for dealing with people, and behaviors that make sense for people get carried over to things.

WHEN WE EXTEND our mental template of love from interpersonal relationships to our relationships with things, some aspects of love, such as enjoying the time we spend with a love object, fit very well. But other aspects of love, such as having an interactive relationship with a love object and feeling a moral obligation to it, are a bit more of a stretch.

Nonetheless, we are able to love things because the brain naturally thinks in metaphors, so our relationships with things only need to be similar to, not identical to, our relationships with people in order to count as love. Part of this similarity comes from the fact that our love for people includes one-way aspects, such as feeling love for people when they aren't around. But this similarity between our love for people and our love for things also comes from our brain's habit of thinking metaphorically. For example, as we saw, musicians' instruments respond to them in ways that the brain, thinking metaphorically, treats as similar to the give-and-take interactions in an interpersonal relationship. And the amount of time we spend using something gets treated metaphorically by the brain as indicative of our closeness with that thing.

I explained in chapter 1 that the brain needs to treat something as at least partly human in order to love it. In chapter 2, I explained that the brain is likely to treat something as human if it inspires anthropomorphic thinking by seeming a little bit human. But anthropomorphism isn't relevant to many of the things people love. Lots of people love things that don't look or sound like people, such as baking, clothing, and beer. In the following chapter, I'll introduce a nonanthropomorphic way that the brain can come to see something as human — or at least human*ish*.

4

People Connectors

*Yes, the violin has its own soul, but it gets that
from another person.*

— Anne Akiko Meyers (b. 1970)

THE VIEUXTEMPS IS A VIOLIN SPECIAL ENOUGH TO HAVE ITS own name. It is 284 years old, and when it was sold for an estimated $16 million, it was the most expensive violin in the world. Its owner has lent it to the renowned violinist Anne Akiko Meyers to use on an ongoing basis. She already had two Stradivarius violins, but as she explained in an interview,[1] "I had to try it and instantly fell in love."

The violin's remarkably colorful sound is of course a big reason she loves it, but she also loves it because it carries the memories of the people who have played it in the past, including Henri Vieuxtemps, the leading Belgian violinist of the nineteenth century, after whom it is named. As Meyers explained, "He was so in love with this violin that he really wanted to be buried with it. . . . Every violin has its own soul, *imprinted by a previous performer* [emphasis added]. . . . I definitely feel the soul of Vieuxtemps with this violin." The violin links two people together in a chain: Anne Akiko Meyers ↔ the Vieux-temps violin ↔ Henri Vieuxtemps. As the renowned consumer researcher Russell Belk[2] would say, Meyers's relationship to the violin is *person-thing-person*.

Things can connect us not just to other individuals but also to groups of people, such as when an heirloom connects you to your family or owning a certain brand connects you to other people who also own that brand. Grammatical issues notwithstanding, as Belk coined the term, person-thing-person phenomena also include these person-thing-*people* situations.

Have you ever noticed that when you buy a new car, you suddenly see that same model of car everywhere? These person-thing-person connections are similar: once you become aware they exist, you realize you've been surrounded by them all along.

Some years ago, I heard a news story[3] that really stuck in my mind. The reporter had been walking through an impoverished shantytown in a developing country. Incongruously, she saw a brand new deluxe yellow leather sofa and love seat standing on the dirt floor of a small ramshackle house. The furniture seemed so out of place that the reporter became curious. As she learned, the home belonged to three brothers whose mother had passed away. The mother had once seen this furniture set in a store window and fallen madly in love with it but didn't have enough money to buy it. After she passed away, her sons pooled their money and bought her beloved furniture as a tribute to their mother. What made the story so memorable for me was that the reporter said something to the effect of "Who would have guessed that the yellow sofa set was actually about their mother?"

The reporter's surprise was entirely understandable. We normally think of our relationships with things as being person-thing, which is to say, a connection between one person and one object. But in keeping with what I call "Belk's first axiom,"[4] if you look closely at your relationships with things you love, you almost always find important person-thing-person connections. In fact, many of our relationships with things are really relationships with people in disguise. This sofa is a case in point.

I call the object or activity in the middle of a person-thing-person chain a "people connector." As to the reporter's rhetorical question "Who would have guessed that the yellow sofa set was actually about their mother?," the answer is this: anyone who has interviewed people about their favorite things. My research on what people love grew out of previous studies that looked at people's special possessions.[5] Researchers conducting these studies discovered that when they asked people to list their most important possessions, or to list objects that were special to them, the research participants mainly listed mementos that connected them to other people. One of the conclusions from that research is that an object's ability to connect people to each other is what usually makes "special possessions" feel special. In one study[6] in which people talked about the art in their homes, the authors summarized their findings by saying that, perhaps surprisingly, no one talked about art in lofty terms, "as the arena in which Praxiteles, Michelangelo, and Rodin wrought their great masterpieces, a hallowed craft to be approached with reverence and a refined sensitivity." Instead, they talked about ways in which the art reminded them of their family and friends.

In my research, I've also found that many beloved objects are people connectors. While showing me the things she loved in her apartment, one woman said:[7]

> These are some things from my grandmother. This [bowl] actually belonged to my great-grandmother. She made it; it's hand-painted. . . . I do think it's rather ugly, but it's something that for years I saw on my grandmother's vanity, so I grew up knowing that it was hers and I associated it with her. So I guess I love that. There are pieces of jewelry of mine that I love. My wedding ring. Stuff that's been gifts from people. Earrings and necklaces, and this is a watch from an old boyfriend of mine. In a sort of strange way, I still love that. Oh, this! Now, there are things here that I really love and again, it's all because they have sentimental value. This [little vase] was a wedding gift. . . . This little Hummel [figurine] was a gift at one time. Most of this stuff is gifts. A few of these pieces are from a very dear friend of mine who died last year. Yeah, I love those things because they have connections with people who are important to me.

Person-thing-person connections also show up in my research when I ask a question about an object but get an answer about a person.[8]

> **ME:** Tell me about your silver cigarette case.
> **PAM:** It really goes back to my dad. . . .

ME: What about your computer?
JOHN: I had this girlfriend who was spending a year in
    Europe. . . .

ME: Let's talk about the antique furniture.
CINDY: My family came from Denmark to America. . . .

ME: Let's talk about your car.
CHRIS: My parents . . .

ME: You also said you loved scrapbooking.
CLAIR: My kids . . .

In another study, researchers Vanitha Swaminathan, Karen Stilley, and Rohini Ahluwalia[9] investigated the question of whether some people are more influenced by a brand's personality than others. Their research showed that when consumers were worried that people found them socially unattractive, their insecurity made them more likely to buy brands that they saw as having attractive personalities. This seems to imply that when people feel insecure about their person-person relationships, they seek to supplement them with attractive person-thing relationships. However, the research also found that a brand's personality only influenced consumers when the product was something that would be seen in public, such as shoes, rather than toothpaste or a clock that is only seen in the privacy of one's own home. If these socially insecure consumers were really trying to create person-thing relationships with the brands as substitutes for people, they would have cared about the brands' personalities even if they were alone with the products.

The fact that the brands' personalities only became important when other people were around showed that consumers were not trying to form relationships with products instead of people. Rather, they were hoping that if people saw them using an attractive brand, this would make them more popular. That is, they were trying to use brands to connect them to other people.

When something functions as a people connector, your feelings about it change. If I like my uncle Morty, for example, I'll tend to like the gift he gave me for my birthday, even if it wasn't exactly what I wanted. If I dislike my aunt Petunia, I'll tend to dislike the coffee mug she picked out especially for me. I call this the *person-thing-person effect,* a phenomenon that occurs when our feelings about an object are driven by the way we feel about the person we associate with it. One man I interviewed[10] told me how proud he was to own a special gold coin that his father had given him. That is, he felt proud of it up until the moment he learned that his father had been having a longtime affair, which led to his parents' divorce. Furious with his father, he not only stopped feeling proud to own the coin, he also gave it away. That's the person-thing-person effect in action.

In chapter 3, I described situations in which people feel they are cheating on brands they love. One of the main things that determines whether people care about cheating on a brand is the extent to which they share a love of that brand with their friends. The more that loving the brand connects them to friends, the more they feel they're cheating if they buy a different brand.[11] The underlying logic here is that if one's relationship with a brand is person-thing-person, buying a competing brand would be disloyal to the other people in the relationship, and this is much more important than being disloyal to the brand.

The fact that cheating on a brand can feel like cheating on your friends brings us to an interesting finding about what types of people are most likely to cheat on brands. You might think that people who are highly materialistic would have intense relationships with the things they own, making them unlikely to cheat on their favorite brands. However, Miranda Goode, Mansur Khamitov, and Matthew Thomson of Western University[12] have found just the opposite: the more materialistic people are, the more likely they are to cheat on brands. This may be because the more materialistic people are, the weaker their interpersonal relationships tend to be.[13] Since people often resist cheating on a brand because it would feel like cheating on their friends, materialistic people, having fewer close friends than nonmaterialistic people do, would have less reason to avoid cheating on brands.

Regarding materialism, it is worth noting that love of money is different from love of most other things. Whereas there is a general tendency for the things we love to connect us to other people, psychologists Xijing Wang and Eva Krumhuber[14] have found that the more someone loves money, the more that person tends to objectify other people, which, as I explained in chapter 1, is antithetical to loving them. So it is possible that, rather than connecting people to one another, loving money makes it less likely that people will form warm, close, and committed relationships with other people.

The overall takeaway is that when it comes to the things we love, the person-thing-person aspects of these relationships are usually very important. But three caveats are needed. First, even when something is loved for its person-thing-person attributes, such as when a mountain climber loves her climbing rope because it — literally and symbolically — connects her to her climbing friends, the relationship can also have person-thing aspects. For example, she might also feel grateful to her rope for saving her on an occasion when she was climbing alone. Second, although the things we love are often people connectors, this isn't necessarily true for the ordinary things in our lives that we relate to in an emotionally cool, pragmatic way. And third, while most person-thing relationships don't involve much love, person-thing relationships with anthropomorphic objects can be exceptions to this rule, because the brain sees these relationships as at least somewhat person-person.

WHY DO SO MANY OF THE THINGS WE LOVE CONNECT US TO OTHER PEOPLE?

There are two main reasons why so many of the things we love are people connectors. First, people connectors can function as relationship warmers. You'll recall that anthropomorphism can be a relationship warmer, meaning that it gets your brain to think of a thing in the emotionally warm ways it normally reserves for people. Creating a strong connection in your mind between a thing and a person can also do that. Suppose you have a strong mental association between a person (your mom) and an object (her favorite teacup). When you think about most teacups, your brain treats them as it would any other thing. But when you think about that special teacup, your brain also starts thinking about a person — your mom. This leads your brain to think about that teacup in some of the same ways it thinks about people, which makes it eligible for love.

The second reason why the things we love are frequently people connectors is that being a people connector makes things deeply meaningful. When old people look back on their lives and say, "I wish I had spent more time focusing on what really matters," they usually mean they wish they had spent more time with their families and close friends. The *people* we feel close to make our lives feel meaningful, while in comparison, most of the *things* in our lives seem trivial and superficial. This doesn't make those things bad; it just makes them ordinary. And our relationships with these ordinary things lack the emotional depth to qualify as loved. By contrast, the things we love stand out by feeling deeply meaningful, and one of the main ways this happens is through their power as people connectors. In fact, when objects are deeply connected to other people, losing an object can feel like losing a person. As one woman who lost her home in a fire explained,[15]

> [I'm most upset about] the memories that I can't get back. Like the pictures that aren't on the internet, or my great-grandmother's blanket that I had from her before she passed away. It really feels like someone passed away.

One of the most unusual examples of the fact that associating an object with a person makes the object meaningful comes from an interview[16] that I think of as "the curious case of the crib that gambled its life away." The interviewee was a man who claimed not to love anything other than people. I was interested in learning if he really didn't love anything or if he just didn't like to use the word *love* to describe his relationships with things.

During the interview, which took place in his home, he showed me a stunning wooden crib that he had built for his baby daughter. When I saw the crib, I was wowed by its intricately patterned inlay and surprised that anyone could have made such a beautiful crib for a daughter and not love it. What prevented him from loving it?

I asked him, "If the crib were transformed into a person, who would it be?" I expected he would describe a nurturing caretaker for his baby daughter. I was so wrong. His answer:

INTERVIEWEE: A female James Bond.
ME: What hobbies would the crib have?
INTERVIEWEE: Backgammon, polo, gambling. I don't
know . . . the crib that gambled its life away.

These answers struck me as so odd that my initial reaction was to wonder if he was joking. But later in the interview I gained an insight that made everything clear. This happened when his wife joined us in the living room, which was filled with art. Most people feel like the art in their home connects them to other people, often to those who gave it to them as a gift. But the man could not remember where any of the artwork came from. His wife then proceeded to tell me in detail about the histories of all the objects in the room, even the ones that had been gifts from the interviewee's friends before he and his wife had met. From what turned into a fairly extended conversation, it became clear that she was the keeper of the relationships in the family, and she saw the artworks as representing those relationships. The man didn't see the artworks as people connectors — he couldn't even remember who gave him what.

His comments about the crib then made sense. I had expected his relationship with the crib to be person-thing-person (father-crib-baby), so some of the love he felt for his baby would rub off on the crib. But he was totally focused on what the crib looked like as an object, so his relationship with it was merely person-thing. From that perspective, given the clean, elegant lines of the crib's design and the black diamond-shaped inlaid border, its personification as a smooth, classy, James Bondesque high-living sophisticate made sense. And it wasn't just the crib. In his mind, none of the objects in his life acted as people connectors. That went a long way toward explaining not just why he didn't love the crib but also why he didn't love any of the things in his life.

THREE TYPES OF PEOPLE CONNECTORS

People connectors mainly operate through three mechanisms: relationship markers, group identity markers, and logistical support.

Relationship Markers

People connectors can connect you to specific individuals (for example, the way a selfie connects you to the other people in the photo) or they can connect you to large groups of people (the way a T-shirt from your high school connects you to everyone else who attended that school). When they connect you to specific individuals, my frequent coauthor Philipp Rauschnabel calls them "relationship markers." Common relationship markers include photos and gifts that mark (remind you of) your relationships with specific people.

When people say that an object has sentimental value, it is usually because it is a relationship marker. A study conducted by marketing professor Marsha Richins[17] compared the feelings generated by three types of objects: cars; recreational objects such as video games and stereos; and typical relationship markers such as heirloom jewelry, mementos, and gifts. This study found that the relationship markers were the most likely to have sentimental value and the most likely to be loved. In some cases, these relationship markers also elicited pangs of loneliness because they reminded their owners of how much they missed the people at the other end of the person-thing-person chain.

The ultraluxury watchmaker Patek Philippe, whose wares cost anywhere from $25,000 to $250,000, knows how powerful relationship markers can be. Given that a $20 digital quartz watch will keep more accurate time than even the most expensive handmade mechanical watch, Patek Philippe needs to give people a reason to buy its products. The company has run an ad campaign for more than twenty years with the tagline "You never actually own a Patek Philippe. You merely look after it for the next generation." These ads position the watches not as practical timekeeping tools but as symbols of generational wealth and family heritage. Positioning the watch as a symbol that represents a father's relationship with his son, or a mother's relationship with her daughter, gives the product an emotional gravitas that helps justify the enormous price tag.

One of the potential drawbacks of relationship markers is that as they accumulate, they can clutter up our homes. Decluttering expert Karen Kingston advises people who have unwanted objects that they can't bring themselves to part with to "accept the love that was given with the gift but let the physical item go." For many people this is good advice, but it's easier said than done. Relationship markers might be deeply infused with the gift giver's love, and it can be hard to separate the two. On the other hand, when people wish to be rid of a relationship, the objects that symbolically marked that relationship are often the first to go.

Group Identity Markers

Even though you may think your identity is just about you, in fact a big part of it is created by your connections to groups. Your penchant for organic food doesn't just express your beliefs about organic farming; it also connects you to other people who buy organic products. The same can be said for lots of things — especially head coverings, including MAGA hats, cowboy hats, hoodies, kaffiyehs, hijabs, yarmulkes, turbans, and other headgear that signals membership in a group.

If the things we love help us form a group identity, what happens to our uniqueness? How might a Cubs fan, for example, retain his or her individuality amid a sea of blue-and-white jerseys? Part of the magic of the things we love is that they sometimes express group belonging and individuality at the same time. Researchers Cindy Chan, Jonah Berger, and Leaf Van Boven[18] point to the fact that all the kids in a high school clique may love a particular style of music that connects them to one another, but each has his or her own favorite musician. Similarly, movie stars might all be dressed to the nines, but each one wears the work of a different designer. It's all part of the way humans signal their connection to, and distinction from, one another.

Loving things has its benefits. For one, we can connect with other people by sharing our passions. HBO takes this so seriously that it builds it into its strategy when it releases a new series. Most other streaming services release a full season of a show at once. This is convenient for fans who want to binge the whole thing on their own schedules. But it also means that when two fans meet, it is likely that one will have completed the entire season while the other may only be halfway through. This makes it hard for fans to connect with one another by talking about the show. To avoid this problem, HBO often releases episodes once per week. That way, most fans will be at the same place in the show at the same time and can talk to one another about what just happened or speculate about what might happen next. These conversations make being a fan a more rewarding experience, so fans keep watching.[19] And for HBO, these conversations constitute free advertising that encourages people who aren't watching to jump on the bandwagon so they, too, can be part of the conversation.

In some cases, entire communities can form around a common passion. Think of Harley-Davidson riders' groups, bird-watching societies, knitting circles, ballroom dance clubs, and countless other groups. This is also true of religion. Many people, especially those who are not religious, overestimate the importance of shared religious beliefs in motivating people to join a congregation. But few people join religious groups because they already believe their doctrines. Instead, they join religious groups because they have begun to find friends and community there. Over time, they take on the beliefs of the group. As the influential early twentieth-century thinker Rabbi Mordecai Kaplan said, "Belonging precedes believing."

A very similar process happens in "brand communities," which are basically fan clubs for a brand. Research has shown that in brand communities, just as in religious congregations, belonging precedes believing. Many people begin exploring brand communities because they have an interest in a brand but don't yet love it. But as friendship with other community members grows, people feel a sense of belonging to the group; this reinforces their love for the brand, which then reinforces their relationships with other people in the community.[20]

The Bronies are a particularly intriguing brand community. Bronies are men (and some women) who deeply love the animated kids' show *My Little Pony: Friendship Is Magic*. No, it has nothing to do with pedophilia; it's just people who are really into the show. And there are a lot of them. *Wired* magazine[21] reported that in 2012, the high point of the show's popularity, there were between 7 million and 12.4 million* self-identified Bronies in the United States. It's a puzzle to figure out what all these guys see in a show squarely targeted at three-to-eight-year-old girls. To hear the Bronies themselves explain it in the documentary *Bronies: The Extremely Unexpected Adult Fans of My Little Pony*, it's just a really great show, so how could they not love it? Let's assume, for the sake of argument, that it is indeed an awesome kids' show. Even so, there has to be more than awesomeness going on.

Here's my take on why Bronies love the show. First, most Bronies range in age from their midteens to their midthirties.

---

* For the record, I'm skeptical of those numbers. But even if it were a mere one million people, that's still a lot.

When I was that age, if a guy said he loved watching a girls' show about magical ponies, he would have been teased mercilessly. But for younger generations, there is social cachet in challenging gender norms. So being a guy interested in a girls' show isn't necessarily considered a shameful thing.

Second, in online discussions Bronies often say that their community includes a lot of young men who are "socially awkward people" (and having fit that description myself, I mean no disrespect). For them, the show functions as a people connector par excellence. And as the title of the show, *Friendship Is Magic*, implies, the stories are all about having friends, accepting other people, and working through conflicts. It makes sense, then, that the show's themes strike a positive chord with this audience. As one Brony, a Northwestern University student, explained in the documentary, "It brings people together. Makes friendships . . . I guess it's cool that a show so ostensibly about friendship is creating friendships in the real world." Another student said, "It's all about the community." A third remarked, "You can share it with other people; that's the magic of *My Little Pony*." *My Little Pony* is person-pony-person.

And third, watching the show is deeply transgressive. As any sociologist of street gangs will tell you, transgressive behavior is a great basis for the formation of tight, mostly male friendship groups. But most Bronies are way too nice to join a street gang.*

---

* Sadly, though most Bronies are good-natured guys looking for friends, there is also a small radical subculture within the Bronies made up of far-right extremists and neo-Nazis. A mass murderer who shot up a FedEx store was a Bronie who was part of this subculture. Figuring out how these people reconcile far-right gun culture with *My Little Pony* exceeds my analytic capacity.

Watching *My Little Pony* is utterly harmless, so it gives them a nice yet transgressive identity that works for them.

When people change their tastes in order to fit in with a group, is this a problem? At one time I would have criticized that kind of behavior as conformist. But now I understand that even though we see ourselves as free and independent individuals, our tastes are strongly influenced by the people around us. When we are forming our tastes in books, movies, music, and other kinds of entertainment, we are aware of what audiences are raving about. For example, upon first hearing a song that our friends love, our brains listen with a task in mind: figure out what's good about the song. Over time, our brains come to recognize certain patterns and features of the "good songs," and we grow to genuinely enjoy them. This isn't the only thing that influences our tastes, but it's an important part of what's going on. Our tastes are a mix of other people's influences and our own unique perspectives. The part that we get from other people helps keep us connected to them, and that's nothing to be ashamed of.

Our tastes can function as walls: they can create social connections to people who are on our side of the wall but separate us from people on the other side. We sometimes love things as much for whom they separate us from as for whom they connect us to. I once interviewed a guy who loved Thai food (this was when Thai food was still new to America) because he felt that it separated him from the "burger and fries" people he grew up with and no longer wanted to have anything to do with. More recently, I was in a bike store when another customer came in and asked the salesperson to help him find a helmet that, in terms of the way it looked, was "the opposite of what those bikers in spandex would wear." "Spandex bikers" are often stereotyped as being affluent liberals. I suspect this customer wanted to stay as far away from that identity as possible, as evidenced by his later saying, "I want a helmet that makes me look like I just fell out of my truck on my way to prison."

Drawing distinctions between our in-group and "the other guys" is a natural part of identity formation and can be based on anything from race and class to sports teams and haircuts. Thinking in terms of in-groups and out-groups seems to be an evolved feature of the human brain, so it's a common way to think about the people around us. But it is possible to control how much emphasis we put on in-group versus out-group membership in our dealings with people. In some societies, opposing groups are killing one another in the streets, while in others, differences between groups are little more than a source of mild pride in one's own group and curiosity about other groups. This is consistent with research showing that it is not just possible but also common for people to love their own group without hating other groups.[22]

As society has gotten richer and more consumerist, in-group versus out-group differences have increasingly been defined by what we own, including the things we love. Once the things we love become markers separating "us" from "them," ugliness can ensue. I am reminded of an example from research conducted by marketing professors Maja Golf Papez and Michael Beverland[23] in which an Apple lover tried to persuade someone not to buy an iPhone because she didn't think the person was the kind of cool, individualistic innovator she believed should own Apple products.

The good news is that this tendency to feel socially distant from people who don't share our tastes can be turned on its head. In my own life, I cringe a bit when I look back at the distaste I felt as a teenager for people who were fans of musical genres that I didn't like. In my post-teenage years, I've explored some of the music that I used to hate so that I could understand what it was about the music that other people liked. It took some time, but as I got to know these musical styles, I started to enjoy them. The beautiful part of this process is that while my old dislike of this music had served as a wall separating me from its fans, once I started enjoying the music, I started to feel a little more connected to the other people who liked it.

Logistical Support

The third way that the things we love function as people connectors is perhaps the most straightforward: our love objects provide practical logistical support for our social relationships — for example, the TV we watch with other people or the boat we go fishing on with friends. Food is particularly good at bringing people together, which is why every wedding is celebrated with a feast. As one woman explained about food's connecting power,[24]

> One of my favorite ways to spend my time is having dinner with a bunch of friends and not just going out to eat somewhere but having a big kitchen where we can all contribute to preparing the dinner, sitting down, and going through several courses — just spending a whole evening around the table, which I guess brings in more than food, but food's, like, the reason we come together.

Nature is probably the most widely loved thing around the globe. The stereotypical image of a nature lover is a lone backpacker trekking through the mountains. That sort of solo adventure can be a profound experience. But more commonly, nature is experienced in small groups, and loving nature is often a person-nature-person affair. Specifically, environmental psychologist Adam Landon and his colleagues[25] found that one of the best predictors of loving natural places is having bonded with other people in those places. Surprisingly, this research also found that just thinking about a natural place we love not only helps us feel connected to nature, it also helps us feel connected to a human community. I suspect that this happens because feeling connected to nature breaks down the mental wall that separates each of us as individuals from the rest of the universe. This gives us an expansive sense of identity that allows us to feel connected both to nature and to people.

Cell phones may well be the most loved commercial product, and the most common reason we love them is that they help us keep in contact with our friends and family. Philipp Rauschnabel and I have found that the more friends people have, the more useful the phone is in facilitating those friendships and the more people love their phones.[26] Following the same reasoning in reverse, the lonelier people are, the less likely they are to love their phones.

SO FAR, WE'VE LOOKED at the first two relationship warmers. Specifically, we've seen that people often enjoy warm emotional relationships with things if the things are anthropomorphic or if they connect us with other people. In the following chapter, I'll begin a discussion of the third, and arguably the most important, of the three relationship warmers. This will be the first of several chapters that look at what it means for the things we love to become part of who we are.

5

You Are What You Love

*In love the paradox occurs that two beings
become one and yet remain two.*

— ERICH FROMM (1900–1980)

PHILOSOPHERS HAVE DEVELOPED COMPLEX MORAL SYSTEMS explaining why it is important to help other people. But if you're hungry, you don't need a philosophical argument to persuade you to eat; it just feels like the obvious thing to do. Similarly, if your children or other people you love are hungry, feeding them also feels like the obvious thing to do, almost as if they were part of yourself. I'll argue here that there's no "almost" about it. In a particular way, the people and things we love do become part of ourselves. But how does that happen? How do objects and activities change from being typical things "out there" in the world to loved things that are part of our identities?

Understanding this has a potential fringe benefit. Almost all of us aspire to be a little (or a lot) better than we are. When we fall in love with something, we make it part of our selves, and in so doing, we transform who we are. Therefore, understanding the processes by which love objects become part of us helps us understand how we became who we are and how we may change in the future.

The idea that when two people fall in love they psychologically merge with each other is one of the earliest documented theories about love. Around 2,500 years ago, Socrates explained love by saying that in the mythological past, humans had four arms, four legs, and two heads but were split in two, and now each one of us must search for our missing half. Or, as Aristotle put it without the colorful reference to mythology, love is "a single soul inhabiting two bodies." Fast-forward a few thousand years, and we find the idea that love is a fusion of identities in the work of prominent psychologists, including Abraham Maslow, Erich Fromm, Theodor Reik, and, most recently, Arthur and Elaine Aron.

Yet when I started researching love as a PhD student, I was deeply skeptical about claims that people and things could become part of our selves. The idea struck me as more poetic than scientific. More important, it seemed to have glaring problems, because there are ways in which even the people and things we love are definitely *not* part of us. Even if we love someone, for example, we can't read his or her mind. And while we can move our hands just by thinking about them, no matter how much we love our dishes it takes more than thinking to get them off the table and into the dishwasher. For these reasons, the idea of love as fusion struck me as the basis of a far-fetched science fiction movie rather than a serious scientific theory. What changed my mind was first getting a clearer sense of what this theory was truly saying, then seeing the considerable scientific evidence in its favor.

WHAT DOES IT MEAN TO MAKE
SOMETHING PART OF WHO YOU ARE?

The idea that things and people could become part of the self seemed wrong to me because when I thought of the self, I equated it with consciousness — the little voice inside our heads that observes what is going on and makes conscious decisions. Religions often view consciousness as part of the soul, which exists separately from the brain and lives on after the body dies. The people and things we love do not — for now, at least (but see chapter 10) — become part of our consciousness.

That said, in addition to consciousness, there is another part of the self called the self-concept, which we often refer to as identity or self-image. When people and things become part of your self, they don't become part of your consciousness, which is why you can't read other people's minds or move dishes just by thinking about them. But they do become part of your self-concept, which makes them part of your self.

Your self-concept refers to all the various ideas you have about who you are. This includes beliefs about yourself, such as "I'm tall" or "I am an artist at heart." It also encompasses important memories that get strung together into a life story. And finally, it includes a very special category that I call the "me category."

Your brain automatically sorts things into categories. If you learn about a new type of Chinese food, for example, your brain puts it in the categories "food" and "Chinese." Your brain also has a me category. Everything that is part of your self-concept, such as your body, your beliefs, and your life story, is included within your me category. When something new becomes part of your self-concept, it simply means that your brain is adding it to the me category.

Imagine a hypothetical person we'll call Jane. The figure below shows the things that are in her me category. The things in a person's me category fall on a continuum, depending on how strongly integrated into that person's self-concept they are.

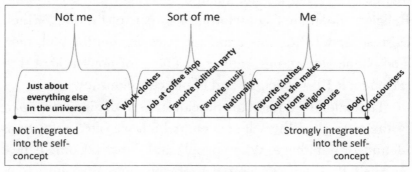

Jane's self-concept

People have an innate belief that their consciousness and their bodies are the most central parts of who they are. Everything else that is part of your self-concept is optional and differs among people. There are many reasons why some of these optional things are more central parts of our identities than others. For example, Jane sees her favorite clothes, but not her work clothes, as part of who she is. This is because Jane works at a coffee shop that has a dress code, so she's not free to express herself through her clothing on the job. But when she's not working, she can wear clothes that she feels are much more "her." Jane also makes quilts as a hobby. She works hard on these quilts and puts a lot of herself into them, so she sees them as more central to her identity than even her favorite clothing. Jane is fairly religious, too, and her religious commitments reflect deep personal values about what it means to be a good person; this makes them central aspects of her identity.

The diagram also shows that there is a "sort of me" category between things that are and are not part of Jane's identity. This sort-of-me zone reflects what psychologists call the "fuzzy boundary" around the self, located on the edges of a person's self-concept, and it isn't always clear whether the things that fall into that category are really part of a person's identity. But just as the existence of twilight does not invalidate the distinction between night and day, the fuzzy boundary around the self does not invalidate the distinction between "me" and "not me."

Your brain treats everything in the me category in quite unusual ways. For example, your brain cares *a lot* about your own safety. And if a thing gets added to the me category, your brain starts to care a lot about its safety, too. The extent to which this happens depends on how strongly integrated into your self-concept it is.

Because the process of making something part of your identity is not conscious, asking yourself "Is such-and-such part of my identity?" provides useful information but is not 100 percent accurate. The *self-referential emotions test* is an interesting way to get a little insight into what is included in your self-concept. Pride, offense, shame, and guilt are self-referential emotions, meaning we only feel them in response to something we have done or when something is said about or done to us. For example, we feel offended when we are insulted, but we don't feel offended when a stranger is insulted. And we feel guilt if we've done something wrong and pride if we've done something well, but we don't feel guilt or pride when strangers do things. Now imagine that your child's teacher tells you that your child's comments in class show flashes of brilliance. Would you feel proud? Absolutely, because you include your kids in your self-concept. Similarly, you feel pride when your team wins, when someone compliments your garden, and when guests are clearly enjoying a dinner you made. The more you feel pride when someone or something is praised and offense when that person or thing is insulted, the more deeply that person or thing is integrated into your self-concept.

The quotation that opened this chapter, Erich Fromm's paradox that, in love, "two beings become one and yet remain two," is correct and not as paradoxical as it sounds. Two beings do become one, because they incorporate each other into their self-concepts. They nevertheless remain two, however, because their conscious minds remain distinct.

EVIDENCE THAT WHAT WE LOVE BECOMES PART OF US

Understanding the difference between the two parts of the self — consciousness and self-concept — makes the idea of love as fusion seem plausible. But just because something could happen doesn't mean it does happen. What scientific evidence is there that, as Helen Keller said, "all that we love deeply becomes a part of us"?

For my dissertation research,[1] I conducted interviews about the things people love. I was interested in psychological theories, particularly the work of Arthur and Elaine Aron, which held that the people we love become part of our selves. I was curious: Would this also be the case for the *things* we love? To avoid "leading the witness," I did not ask people directly if the things they loved were part of their self-concepts. Rather, I asked them what made these things so special.

To my surprise, without any prompting, many people described their love for things as a kind of fusion. As one person explained,

> [The things you love are] part of your identity,
> what you think of yourself; there's no separation.
> When you're talking about what you love . . . you
> are essentially talking about yourself.

But these interviews also showed, in a roundabout way, that one's conscious mind doesn't always know if a particular thing is part of one's identity. I ended each interview by asking people if they had gotten any insights about themselves or the things they loved through the process. It was fairly common for them to say that before the interview, they would not have described a particular love object, such as a cell phone, as being part of themselves. But just by talking in depth about their love objects, they realized that those objects were part of who they were. Even though this supports the idea of love as fusion, it also shows that before the interview, *the interviewees' conscious minds were wrong* when they thought their love objects were not part of their selves.

Even if their conscious minds didn't always know whether their love objects were part of their identities, there were lots of clues throughout these interviews that people were thinking about the things they loved as being aspects of themselves. One of the most interesting clues was the way people would decide if they loved something: they would often ask themselves, "Could I live without it?" One person made clear that this question wasn't literally about dying when she said that she didn't truly love food because she "could live without it."

A comment from a different interview[2] gave an interesting clue about what the phrase "can't live without it" is really getting at.

> Writing is a greater part of my existence and who I am than war-gaming is. Without writing, I would die; without war-gaming, I would live.

Without writing, he "would die," not because the loss of writing would hurt him physically but because it is so central to his identity. This next quotation, from a woman who only sort of loves popcorn but truly loves music, clarifies the issue.

> I wouldn't be any less of a person if I didn't eat popcorn. [But] without music, I would be dead. . . . I would have to be an entirely different person if I didn't have it.

Notice how this woman equated being dead with becoming an entirely different person. If you lose something you love, it stops being part of your identity, so the person you were before no longer exists. Therefore, "you" cannot live without it.

Another source of evidence for love as fusion comes from what are called "projective questions," such as "If the love object were magically transformed into a person, who would it be?" These questions are good at getting past a person's conscious beliefs and uncovering deeper thoughts and feelings. When describing who their love objects would be, people have a strong tendency to imagine the things they love as mirror images of themselves. For example, one person personified his cat as someone sharing his taste in music, books, travel, hobbies, and politics — even his tendency to be set in his opinions! (Have you ever met a cat that wasn't set in its opinions?) One woman responded to this question by saying that both her plants and music would be women who were thousands of years old. Since she was a mere thirty-seven years old at the time, at first this seemed like an exception to the common mirror-image pattern. But she surprised me later in the interview by saying that she felt similar to her plants and music in this respect because "I feel as though I am a very old soul, and so are they."

There is also evidence from experimental psychology that supports the idea that love is fusion. Arthur and Elaine Aron were the first experimental psychologists to extensively investigate whether people included their loved ones in their identities. A lot of their research employs the same basic strategy: find something that is true about the self but not true about other people (e.g., we feel pride when we are praised, but not when other people are praised). Then see if it also applies to the people we love (e.g., if we feel pride when the people we love are praised, this shows that we are implicitly seeing them as part of our selves).

In one study, Arthur Aron and his colleagues[3] gave a group of students between $10 and $20 and told them they could split it any way they liked between themselves and a stranger. Not surprisingly, these students tended to keep almost all of it for themselves. But another group of students was told to divide the money between themselves and a close friend who lived out of town. The students were told that the researchers would mail the friend a check, but the friend wouldn't be told how much money the student had been given to split up. So the friend wouldn't know how much of the money the student had kept for himself or herself. These students usually divided the money evenly, but in a few cases they gave their friends more than half. This simple study shows that as relationships become closer, we treat others as we treat ourselves.

You've probably noticed that when we perform poorly, we often make excuses for ourselves by blaming something or someone else for the problem,[4] but we are much less likely to make those types of excuses for other people's failures. Marketing professor C. Whan Park and his colleagues[5] have shown that this happens with the things we love, too. Specifically, when something goes wrong with a product you love, you tend to make excuses for the product and say it was just a fluke of some kind. So when we love a product, we treat it as part of our selves.

Other studies are particularly interesting because they show that integrating the people and things we love into our selves takes place at a deep and nonconscious level that affects the basic workings of the brain. For example, a study conducted by social psychologists Sara Konrath and Michael Ross[6] noted the quirky fact that people tend to remember their past successes as being more recent than they actually were and their past failures as occurring longer ago than was truly the case. However, we don't do this for the successes and failures of strangers. Yet when we think about the past successes and failures of people we love, we have the same quirky memory bias that occurs when we think about ourselves.

Recent evidence from neuroscience provides further support for the idea of love as fusion. Neuroscientists have known for some time that there are certain areas of the brain that are activated when people think about themselves but not when they think about other people. Researchers Shinya Watanuki and Hiroyuki Akama[7] reviewed all the relevant studies and found that when we think about people or things we love, our brains show the same patterns of activation as they do when we think about ourselves.

Considering all this evidence, I can only conclude that the idea that love involves fusion is poetic, romantic, and true. Yet there is also a bit more to it than that.

SOME WAYS WE BECOME WHO WE ARE

Including something in your self-concept requires time and effort on the part of your brain. There are a number of mental processes by which this takes place.

Thinking and Learning About the Love Object

It's common to be a little obsessed with the things you love, especially when the relationship is new. This obsession plays a functional role in integrating the love object into your identity. All that obsessive thinking about the thing you love reinforces its place within your me category, making it part of who you are.

In my research with Rajeev Batra and Rick Bagozzi,[8] we have found that having frequent thoughts about something that just "pops into your mind" is an important part of loving things.

This was also demonstrated in previous studies about interpersonal love. For example, research[9] has found that in dating couples, the amount of love that people feel for their partners is even more influenced by having frequent thoughts about them than it is by finding their partners good-looking or by having intimate conversations together.

"My neighbor got a new car today and he's been standing outside staring at it while eating chips for 15 minutes"

Another gift from the internet. Get a garage, you two!

Loving something is connected not only to thinking a lot about it but also to knowing a lot about it: characters in romantic books and movies will sometimes ask, "How can you say you love me when you don't know anything about me?" The importance of having deep knowledge about the things we love is particularly striking in the comments of one woman I interviewed.[10] Originally, she had claimed to love music, but upon further reflection she changed her mind, saying she only liked it.

> I like listening to music, [but] when I think about
> it in terms of putting it in the category of love, no

[I don't love it]. I have difficulty with music. I can't recognize instruments very well. Another thing about music that I'm not very good at is identifying [what band plays what song]. That's my problem with music — I don't understand it.

Despite her enjoyment of music, her lack of expertise created a psychological distance between her and music, which was incompatible with love.

Why is knowing a lot about something connected to love? The existentialist philosopher Jean-Paul Sartre said that knowing something deeply was a way of integrating it into your self-concept. But for this to work, the knowledge must be passionate and intimate rather than detached or coldly intellectual. When we know something in a passionate and intimate way, we come to care about it rather than simply seeing it as an object that exists only to meet our needs. In Sartre's view, then, it is no coincidence that sexual relations are sometimes referred to as "knowing" the other person.

Including Love Objects in Our Life Stories

Pamela Paul, a writer and the editor of the *New York Times Book Review*, was interviewing the author and legal analyst Jeffrey Toobin[11] when she mentioned that they had something unusual in common. They both kept "a book of books," a notebook in which they tracked every book they ever read. It turns out they had something else in common, too.

JEFFREY TOOBIN: My dad died, unfortunately, very much on the young side. . . . I found among his

possessions this little leather notebook. And I saw
that he had written in it every book that he had
read as he read it. And what's especially poignant
is — my father had a brain tumor, and you can
see his writing deteriorate as he got sicker and
sicker. I kept this wonderful keepsake of my dad,
and then three years later I thought to myself,
what the hell, I'll just start doing the same thing in
the same book. And here it is many decades later
that I still have the same book. . . . It is my most
cherished physical possession. There was a fire in
our apartment building on a couple different floors,
and my wife and our two kids, we had to evacuate,
and the only thing I took besides our cat and our
kids was my book of books.

PAMELA PAUL: That's so funny because I start my
memoir with that exact anecdote. Luckily I have
not yet had that fire, hopefully I won't, but if I did,
that's the thing I would grab . . . after the kids.

What is it about these books of books that makes them so
meaningful and important to both writers? As we go through
life, we create a mental story about the past, present, and future
that forms a central part of our self-concepts. The things we
love can become part of our identities by playing a role in that
mental story. I see this a lot when I meet people who love dia-
ries, souvenirs, and memorabilia that mark important times in
their lives. It is also common to integrate things into our life stories that function as trophies for major accomplishments, such

as first houses and diplomas. For example, a freelance journalist once told me that she loves seeing her name in print: "It makes me feel great about myself. It's concrete proof of what I have just accomplished."[12] Another person I interviewed once said about a car he loved,[13]

> It's not so much the car but what the car represents. Because all the while I was trying to get through school, I saw the car, and I wanted it. It represents dreams coming true for me. Getting through school was my first goal that I really attained, but getting the car made me feel like I could do anything.

Integrating the things we love into our life stories is not limited to objects from the past. Fantasy possessions can serve as imagined trophies for future accomplishments. I once interviewed a woman who grew up poor and had only a high school education but had worked her way up to become a manager at a beauty salon and had ambitions to go further. At the beginning of the interview, she said it wasn't possible to love material objects. But later, she admitted that she loved the white-on-white Jaguar she hoped to own someday because, as she said with a laugh, "it's chic, elegant, and it runs and purrs like a kitten, just like me."[14] Why did she feel strongly enough about this car to say she loved it after having said that people can't love material objects? For her, there was no great contradiction between saying one can't really love a material object and saying she genuinely loved this car, for in her mind, the car was not primarily a material object but rather a symbol of her aspirations for herself.

Physical Contact

Along with thinking about love objects, having physical contact with them causes them to rub off on us (metaphorically, if not literally) and become part of who we are. When it comes to items once owned by celebrities, for example, the more physical contact the celebrity had with the object, the higher its value. Memorably, a tape measure once owned and frequently used by Jacqueline Kennedy Onassis sold for $48,875 at auction.[15] In the same spirit but in the opposite direction, physical contact between an object and negative celebrities, or infamous villains, reduces the object's value and makes people inclined to throw it away if they own it. In fact, researchers Carol Nemeroff and Paul Rozin[16] found that people refused to try on a sweater if they were told it had once been worn by Adolf Hitler. This brings to mind a woman who complained on Reddit that her fiancé's mother had tried on her wedding dress without permission and that just knowing this made her want to get a new dress for the wedding.

This fact — that things we touch rub off on us — has interesting implications for clothing stores. Marketing professors Jennifer Argo, Darren Dahl, and Andrea Morales[17] found that people are less likely to purchase an item if they think about the fact that it has been tried on by another customer but more likely to purchase something that has been handled by an attractive salesperson of the opposite sex. However, if a previously used item has an interesting story, especially one that establishes an emotional relationship with a customer, it can increase its desirability. The Oxfam resale shop in Manchester, England, has turned this phenomenon to its advantage by asking people who donate goods to share the story of the objects they donate and what those objects meant to them. As part of Oxfam's Shelflife project, this information is then made available to shoppers through an app that reads a barcode on the item and pulls up the item's story, such as:

> Well my item is the little red silk make up toiletries bag. . . . It's from a place called Narai in Bangkok and it was one of the very first things that I bought when I went to visit my uncle and his wife Noi who lived just outside Bangkok themselves and I believe if this is the shopping trip that I'm thinking of, I believe it's also one of the very first times that I got a tuk tuk [motorized rickshaw] and nearly fell out, on the middle of the motorway, on the way back which I'm pretty certain it is actually so yeah that's my story and I risked life and limb to get that toiletries bag.

As Oxfam's Sarah Farquhar explained, "Items with an interesting story behind them are instantly more appealing to our customers." This is attractive to Oxfam not just because of the revenue it brings in but also because getting people to reuse and maintain objects instead of buying new ones encourages a green economy. Telling the stories behind products helps "prevent them [from] heading for the landfill" and "will encourage people to love items for longer."[18]

Have you ever wondered why there are women's and men's shampoos but there aren't women's and men's laundry detergents? There is nothing about women's and men's hair that requires different formulations of shampoo any more than there's something about men's and women's jeans that requires different formulations of detergent. Shampoos are marketed as masculine or feminine because they are applied directly to the body, which gives them a strong symbolic relationship to one's sense of identity. Marketers know this, so the scent, packaging, and advertising of toiletries are all geared toward imbuing products with certain qualities — such as sexiness, naturalness, sophistication, wholesomeness, masculinity, femininity, and what have you. Many consumers are willing to pay extra for products that match their desired identities, including their gender identities, and marketers both amplify and capitalize on this tendency.

If rubbing a shampoo on one's head leads people to integrate it into their selves, it's not surprising that eating something can have an even stronger impact. The saying "You are what you eat" seems to refer to the fact that food becomes part of your body, but it can also refer to the way the foods we eat, and refuse to eat, can become part of our identities. Physically, people only digest the things they *do* eat; but on a symbolic level they can also incorporate their refusal of certain foods into their self-concepts. That's why, for example, when people decide not to eat meat, "being a vegetarian" becomes an important part of their identities. It is also why so many religions restrict the faithful from eating certain foods: Hindus don't eat beef; Jews don't eat pork or shellfish; Muslims don't eat pork; and Jains don't eat meat or vegetables that grow underground. These food rules are a powerful way to strengthen the integration of a religion into an adherent's identity.

Control

Some things, such as tools, cars, musical instruments, and sports equipment, can become part of our selves when we have intuitive control over them. For example, once we are experienced drivers, if we want to make a turn, we don't think, *Step gently on the brake to slow down and turn the steering wheel to the right*; we just think, *I want to be over there*, and our bodies automatically do what needs to be done. This kind of intuitive control over a car is similar to the control we have over our bodies. So we come to see objects that we can control in this way as extensions of our bodies and hence as parts of our selves.

Creation and Investment

It has been said that myths aren't things that are false because they never happened; they are things that are true because they keep happening over and over again. This certainly applies to the Greek myth of Pygmalion, in which a sculptor creates a statue of a woman so beautiful that he falls in love with it. The pattern in this mythic story repeats itself continually as people fall in love with things they make.

We tend to see the things we create as being parts of ourselves because, like our children, they spring from us. Or, as is sometimes said, *we put a lot of ourselves* into the things we make. As Erich Fromm writes in his classic book *The Art of Loving*,[19]

> In any kind of creative work the creating person unites himself with his material, which represents the world outside of himself. Whether a carpenter makes a table, or a goldsmith a piece of jewelry, whether the peasant grows his corn, or the painter paints a picture, in all types of creative work the worker and his object become one, man unites himself with the world in the process of creation.

Each of us is, to ourselves, of immense value. So when things become part of our selves, we come to see them as valuable, too. Research shows that people value things a lot more when they have helped design or build them. In a 2012 study, researchers Michael Norton, Daniel Mochon, and Dan Ariely[20] asked people to create origami figures. The researchers then asked the participants how much they would be willing to pay for the figures they had just created and how much they expected that other people would be willing to pay for those same figures. Then the researchers asked a comparable group of people how much they would, in fact, pay for the origami figures the study participants had made. People said they would pay almost five times as much for an origami figure they made themselves as a stranger said they would pay for that same figure. Perhaps more surprisingly, when the origami makers were asked what they thought other people would be willing to pay for their figures, they assumed that others would spend around the same amount as (or in some cases even more than) they themselves would pay.

I've found that people are most inclined to see something as part of their selves when they invest their creativity in it. But any effort, even if it's not very creative, can have this effect. For example, business professor Peter Bloch[21] found that people who washed and maintained their cars themselves were particularly likely to see their cars as part of their identities, and Russell Belk[22] found the same thing to be true about caring for one's home. Similarly, Norton, Mochon, and Ariely[23] showed that even just assembling things according to directions leads people to place an irrationally high value on objects. They call this "the IKEA effect," after the retailer of ready-to-assemble furniture.

It Changes You

I've just proposed that when you change an object, it can become part of your self-concept. This also works when the object changes you. People especially tend to see something as part of themselves when it changes them by helping them expand or grow as a person. For example, I've interviewed men who love working out and who see exercise as part of who they are because of the way it changes their bodies. Similarly, a book lover I interviewed said that books are part of him because, as he expressed it, "you incorporate them in such a way that it just adds on and on and on about how you would look at life; it's sort of expansive for myself."[24]

Arthur and Elaine Aron have studied how falling in love with a person changes who you are, and their findings are also relevant to the things we love. The Arons argue that humans have an innate drive to grow and develop. They see falling in love as connected to this drive for personal growth because it allows us to include beloved people within our selves. In one study they conducted with psychologist Meg Paris,[25] they asked students to answer the question "Who are you today?" every two weeks by listing the words they felt described themselves at that point in time. The researchers concluded that these self-descriptive words represented nineteen aspects of identity. They included words that describe emotions (e.g., *angry* and *happy*), family roles (e.g., *son, daughter, brother, sister*), occupations, and so on. Each time the students prepared their lists of self-descriptive words, the researchers asked them if they had fallen in love over the previous two weeks, and because the participants were undergraduates, falling in love wasn't all that uncommon. Students who had just fallen in love described themselves using words from a significantly wider variety of categories, showing that falling in love had expanded and broadened their self-concepts.

This holds true for things, too. A friend told me that he had surprised himself by discovering that he loved watching documentaries about wild animals, a genre he'd always considered too dull to bother with. He said, "I never thought of myself as someone who watched nature shows, but I guess I am." His self-concept expanded when his new passion revealed an aspect of himself that he hadn't known was there.

Buying and Owning

More than one hundred years ago, William James, who is often considered the first modern research psychologist, pointed out the strong connection between owning something and seeing it as part of your identity: "Between what a man calls *me* and what he simply calls *mine* the line is difficult to draw. We feel and act about certain things that are ours very much as we feel about ourselves."[26] James further observed that we see everything that is part of our identities as one of our *possessions*: *my* body, *my* memories, *my* country, *my* religion, *my* favorite band, and so on. That's why love involves so many phrases like "Will you be mine?"

There is a distinction between legal ownership and psychological ownership. For example, we have psychological ownership of a country that we refer to as "my country," but we can't charge everyone who lives there rent. And students often feel psychological ownership over their regular seats in a classroom even though they don't legally own them. When it comes to integrating things into our identities, it's psychological ownership that really counts. But legal ownership is still important, because legal ownership usually augments psychological ownership.

People who start a new activity often buy all sorts of expensive equipment and clothing that strengthens their sense of psychological ownership of the activity. This can turn out to be a costly mistake. Even people who have loved an activity for a long time can overemphasize buying things as a way of engaging with that passion. I was once at my local bike shop getting my bicycle fixed when I got into a conversation with another customer who was thinking of buying a $7,000 mountain bike. He already owned eight mountain bikes and said he had trouble storing them all; if he brought another one home, his wife was going to kill him. He bought the bike anyway. This guy may have suffered from a compulsive-consumption and/or hoarding disorder (and even if he didn't, I'll bet his wife thought he did). But it's common to see less pathological examples in which people express their love for an activity by buying all sorts of paraphernalia that they never use.

I've wrestled with this issue myself, so I know whereof I speak. Since high school, I've been a bit of an audiophile. Unfortunately, as a friend of mine once noted, "audiophiles are never satisfied," and stereo equipment can be very expensive. Worse yet, I find that when I get too engaged with the technology, I stop really listening to the music because I'm distracted by my thoughts about my stereo. I've found that a good way to keep my audiophile tendencies in check is to consciously focus on aspects of the music that aren't about the accessories. For example, I avoid reading about audio equipment and turn my attention to reading about music instead.

Boundary-Breaking Experiences

It's a familiar movie plot: strangers are thrown together on a harrowing adventure from which they emerge as friends or lovers. You may have noticed this in real life as well: it's often the little disasters during vacations with friends, rather than the sunny days spent sitting on the beach, that provide the basis for real bonding. Adam Sandler even has a bit in one of his comedy routines in which he talks about "falling in love" with a guy whom he happened to sit next to on a very intense roller-coaster ride because there was something about going through that experience together that created a bond between him and the stranger. Arthur and Elaine Aron call these kinds of stressful adventures "boundary-breaking experiences" because they break down the mental boundaries between people and make it easier for another person to become part of one's self-concept.

Something similar happens when we go through intense emotional experiences with the things we come to love. In my research, people often talk about loving music, books, and other things that helped them get through emotionally difficult periods in their lives. These people express a sense of gratitude toward the things that helped them weather the storm, and they feel strongly bonded to these love objects.

These bonding experiences sometimes involve fear. I once spoke to a young man who loved horror movies, and he changed the way I thought about them. I asked him why he enjoyed being scared. He replied, "I don't enjoy just being scared; I enjoy being scared with my friends." It's the shared bonding experience, not the fear itself, that made him love those movies.

Lea Dunn, of the University of Washington, and JoAndrea Hoegg, of the University of British Columbia,[27] have shown that not only do scary movies help people bond with their friends, they also help people bond with brands. Dunn and Hoegg told study participants that they would be evaluating both a new brand of sparkling water and some movie clips. They then gave each participant a glass of the sparkling water and played movie scenes that were either sad, scary, happy, or exciting. The viewers then rated how emotionally attached they were to the brand of sparkling water they had just tried. Those who had seen movie clips that were sad, happy, or exciting all had low levels of emotional attachment to the brand of water. But the people who had seen the scary film clips reported dramatically higher levels of emotional attachment to the sparkling-water brand. Why did this happen? The researchers believe that our brains evolved so that if we're in a scary situation, we bond with the people around us, because having a friend in tough times can be especially useful. This mental mechanism carries over to our relationships with things.

These boundary-breaking experiences can also help explain why sports fans often love their teams so much. Sports psychologist Sarah Broadbent[28] reports that sports generate stronger emotions than any other type of spectator entertainment. Each game is an emotionally intense, and often scary, boundary-breaking experience that fans go through with a team. This helps make the team a big part of fans' self-concepts, which is why they say "We won" or "We lost" even though they never got up off the couch. I recall seeing a cartoon in which a wife says to her husband as he watches football on TV, "Sometimes I think you love football more than you love me." To which the husband replies, "Well, at least I love you more than I love hockey." But that's just a cartoon . . . right?

As YOU'LL RECALL, relationship warmers are the mental mechanisms that change your normally cool, pragmatic relationships with things into warm, emotionally attached relationships. At this point I've touched on all three relationship warmers: anthropomorphism disguises things as people; people connectors create a series of mental connections between you, the love object, and the other people in your life; and incorporation into the self makes things part of a very important person — you. Of the three, incorporating things into your self-concept is the most prevalent and most significant, and I'll soon explain why. The following three chapters will cover various ways in which the things we love shape who we are.

6

Finding Ourselves in the Things We Love

Love is but the discovery of ourselves in others,
and the delight in the recognition.

— ALEXANDER SMITH (1829–1867)

THE YEAR WAS 1950, AND FOR NESTLÉ, SALES OF NESCAFÉ instant coffee in the United States were disappointing. When the company asked American women why they didn't buy Nescafé, the overwhelming answer was the flavor. But blind taste tests showed that in fact, compared to the other lousy coffee most people drank in 1950, the taste of Nescafé was fine. What was really going on?

Nestlé asked the consumer psychologist Mason Haire to find out.[1] He discovered that women who rejected Nescafé almost always saw their behavior as entirely rational, telling themselves, "I avoid instant coffee because it tastes bad." But by digging deeper behind their feelings, Haire discovered that these women looked down on instant-coffee buyers as lazy, spendthrift, disorganized, and, worst of all, in bad marriages. Presumably, women in happy marriages took the time to brew their husbands a nice cup of "real" coffee. As consumers, we tend to mistakenly believe that we only like, or dislike, things (such as coffee) for straightforward reasons (it tastes bad). But *identity issues* — the desire to be a certain type of person (in this case, not to be a lazy, spendthrift, disorganized marital failure) — tend to play a much bigger role in what we buy, and what we love, than we realize.

When I started studying consumer behavior, I was surprised by how frequently researchers would try to explain people's tastes by looking at the types of identities they want to have ("She drives a pickup to show she's a patriotic American conservative, whereas he drives an electric car to show he's an ecologically minded progressive"). I recognized that identity matters to people, but it seemed to me that some researchers were a little obsessed. Later, as I started doing my own research, I discovered that most researchers aren't overly focused on the topic of identity; they're just reporting what actually motivates consumers, even if consumers themselves are only partly aware of it. When I learned more about the history of Western culture, that all started to make sense.

THE ROMANTIC LOVE OF THINGS

Things play a much bigger role in our lives today than they did at any time in the past. The changes that led to the current state of affairs began around 1760, with the Industrial Revolution, which started in the textile industry and led to dramatically lower prices for clothing before eventually leading to lower prices on almost everything. To give you some perspective on this, during the Industrial Revolution the price of fabric fell faster than the price of computer power did during the digital revolution. The economist Adam Smith, who wrote about the Industrial Revolution as it was happening, remarked that owning a linen shirt, an item once reserved for the affluent, had become a social necessity for even the lowest-paid laborers.[2] As the Industrial Revolution progressed, houses started to fill up with possessions, and stuff came to play an increasingly large role in people's lives. This happened among the upper classes earlier than it did among others, but stuff now permeates the lives of all but the poorest people around the world. When I hear the cliché about Inuits having many different words for snow, I sometimes picture an Inuit pondering why we have so many words for various kinds of clothing, cars, appliances, and stores.

But the changes brought on by the Industrial Revolution went far beyond simply inundating us with stuff. The increasing wealth also led to a profound cultural change — gradually increasing our individualism. Why? Individualism means that if we face a choice between doing what our internal desires tell us to do versus what our families, friends, societies, religions, or traditions tell us to do, we will follow our internal desires. And the more money we have, the less power other people have over us. So having money makes it easier to be individualistic — to ignore what other people want us to do and instead follow our internal desires. Evidence for this comes from many studies showing that (1) within any given society, rich people tend to be more individualistic than poor people,[3] (2) rich countries tend to be more individualistic than poor countries,[4] and (3) if a country becomes richer over time, it also tends to become more individualistic.[5]

This change is very powerful but also very gradual. It has taken Europe and North America hundreds of years to become the individualistic cultures they are today. In Europe, as the Industrial Revolution led to economic growth, the beginning of the shift toward individualism took the form of a social movement called Romanticism. This movement argued that each of us has an inner "authentic self,"* that we should "find ourselves" (i.e., figure out who that self is), and that we should live a life that is true to that inner self even if it requires challenging our parents, community, or social conventions. This movement became so influential that historians refer to the period circa 1800–1850 as the Romantic Era.

This Romantic idea of the authentic inner self had a profound impact on marriage. Instead of following their parents' views about whom they should marry, young Romantics argued that love is the voice of the inner self guiding them to the correct marriage partners. Because the idea that love should determine whom you marry became widely popular in the Romantic Era, the love between a courting couple became known as — you guessed it — romantic love. But this notion of being true to your inner self went far beyond marriage.

Consider the things that have a big impact on identity, such as where we live, who our friends are, our jobs, our religions, our genders, whom we love, how we dress, and whom

---

* Henceforth, I will use the term "authentic self" because that's the common name for this idea. But I don't mean to imply that I agree with the Romantic notion that each of us has a single, authentic inner self that we should listen to at all times. Reality is more complicated than that.

we marry. Before the Romantic Era, everything on that list was largely out of an individual's control. The vast majority of people followed narrow gender roles, lived near the places where they were born, and followed their communities' traditions in religion, food, music, and clothing. They also went into their families' occupations and married people whom their parents approved of or even chose for them. But today, especially in Western societies (and increasingly in non-Western societies), all these aspects of our lives are choices we get to (and have to) make.

Not only did the Romantic movement allow us to make these decisions, it also developed a philosophy of *how* we should make them. Romanticism advocates being true to our inner selves. Therefore, when I talk about the romantic love of things, I don't mean a sexual or even passionate love; I mean loving something because you see it as reflecting your inner self.

Many of us are so steeped in Romantic ideas about love that it may not occur to us that there might be other ways of thinking about it. Isn't love always a matter of following your inner desire for a person or thing? Not for everyone. I recall interviewing a woman in Singapore who told me she loved a dress she had worn to a big social event and felt it was really "her," *even though she didn't like the way it looked.*[6] She explained that loving and helping her family were much more important parts of her identity than was her taste in clothing. At the big social event, her mom's friends told her mom how nice her daughter looked and what good taste she had. This brought honor to her mom. As this woman saw it, her authentic self was being a person who loved and honored her parents and brought respect to her family. This dress helped her enact that aspect of her identity. Therefore, the dress was "her," regardless of her inner feelings about the way it looked.

This woman is a perfect example of what researchers Hazel Markus and Shinobu Kitayama[7] call the "interdependent self-concept." Their work contains abundant evidence that it is common in collectivist cultures to see one's authentic identity as being defined by one's connections to other people (who your family is, who your friends are, where you work, and so on) rather than by one's individual tastes and beliefs. I don't doubt that's true. But through interviews I've conducted in economically successful Asian cultures,[8] I've come to suspect that Romantic, individualistic ideas about love have become increasingly popular. That's partly attributable to the growing wealth in these cultures, which brings individualism along with it, as it does everywhere else. It may also partly be the result of Western cultural influences.

Although historians often date the decline of the Romantic Era to around 1850, the core idea about being true to one's inner self never went into decline. In every era that followed, the belief in an authentic inner self continued to gain strength. For example, feminism is rooted in this same basic idea: women should look to their authentic inner selves to discern how they want to live, even if this conflicts with traditional ideas about women's roles. Today, the LGBTQ+ movement has found great success by encouraging people to follow the dictates of their own sexual identities rather than traditional ideas about what is acceptable. Over the past 250 years, the disruptive power of finding and following one's inner self has been demonstrated time and time again. As the cliché goes, "kids these days" are rejecting tradition and are just doing whatever they want.

THE PROJECT OF THE SELF

The growing popularity of the idea that people should live according to the dictates of their inner selves is, on the whole, a good thing (although, like any good thing, it can be taken too far). People living in societies that allow individuals to choose their careers, marriage partners, religious practices, and political affiliations tend to be happier than people living in more traditional societies, where these types of decisions are largely dictated by tradition, family, or other circumstances of birth.[9]

But having the freedom to live according to our inner selves also encumbers us with a huge life project. It's hard enough to make a decision about who you want to be, but that's just the first step of this project. Next you have to make that vision a reality — i.e., turn your inner self into your *actual* self, which is to say, you have to *self-actualize*. This becomes a central effort in one's life, a *project of the self*.

Because in contemporary individualistic cultures, almost everything we do is supposed to express our inner selves, the project of the self pervades all aspects of our lives. For example, when you do something as mundane as buying a shirt, you don't just ask if it fits your body; you also ask if it fits your identity: "Is it me?" Life becomes a stream of options to change or fine-tune your identity: Should I change jobs? Move to a new home? Start going to church? Stop going to church? Buy a pickup truck or maybe an electric bike? Take up a new hobby? Become a vegan? The choices are endless.

Not only does the project of the self touch every aspect of our lives, it also continues for the entirety of our lives. In the 1950s, the psychologist Erik Erikson noticed that, as Western culture increasingly emphasized being true to your inner self, adolescents were finding this to be a difficult task. So he coined the term "identity crisis" to describe what they were going through. Today, middle-aged folks discover that, although identity concerns decrease with adulthood, they don't go away. Even when people finally reach retirement, they often see it as yet another opportunity to reinvent themselves.

So here we are, modern humans facing a need to discern and create our identities while surrounded by a sea of things to own and do. Being clever animals, we put the pieces together and use those things to help define our identities. This is why the need to define our identities has such a big impact on what we like.

A MAGIC COMPASS

In the *Pirates of the Caribbean* movies, Captain Jack has a magic compass that points toward whatever the person holding it wants most in the world. For all of us who are engaged in the project of the self, discerning our true identities is one of the things we want most. When we fall in love with something, that experience shows us a little more of who we are. In this way, the things we love are like a magic compass pointing us toward one of the things we want most in the world — knowledge of who we really are.

I said in chapter 5 that when we create something, we are *putting* ourselves into it. By contrast, when we try something and discover that we love it, we are *finding* ourselves in it. Before Maya Shankar became a leading cognitive scientist, she had a promising career ahead of her as a classical violinist. In an interview,[10] she explained, "One of the great blessings of playing the violin is that it allowed me to see what it really felt like to be in love with something and to be really passionate about something." That kind of passionate love allowed her to find herself through the loved activity. As she put it, "You see . . . features and traits that are extracted [from you by] engaging with that pursuit." Tragically, her dreams of being a violinist were ended by an injury to her left hand. But she knew what it felt like to love what you did, so she started looking for something new that would ignite her passion and bring out the same qualities in herself that she had discovered through the violin. As she put it, "Your hope is that in the new explorations [of things you might love], those [new activities] can extract those same qualities from you." She fell in love again, this time with research in cognitive science. She has since become something of a superstar, serving as the senior adviser for the social and behavioral sciences at the White House Office of Science and Technology Policy under President Obama before going on to become the first head of behavioral insights at Google.

INTUITIVE FIT AND LOVE AT FIRST SIGHT

People often feel a sense of intuitive fit, or "rightness," about the things they love that can feel like love at first sight. You may recall the woman in chapter 1 who fell instantly in love with a potential new home. Another woman told me about a similar experience.[11]

> I remember when I was really young, [Victoria's Secret] had an Aquarius body lotion, and this is

one of my first memories of ever setting foot in there. And I thought, *I'm an Aquarius*; out of all the smells of all the different signs, that was the one that I loved the most, too. I knew that before even seeing it was Aquarius. I'm like, *This was for me, and they knew it.*

In a milder form, we frequently have these experiences when shopping. We look at, say, shoes on display and dismiss most of them with barely a glance. But one or two pairs seem to leap out at us, urge us to try them on, because they just seem "right." That sense of rightness about an object makes us feel that it is authentically part of who we are — that in seeing the object, a part of us has become visible.

INTRINSIC REWARDS AND CHEMISTRY

Imagine you are out on a date with someone who is honest and professionally successful, but you don't enjoy the time you spend with that person. In this case, even if you respect your date, you wouldn't say you love him or her; there isn't any "chemistry."

We have a similar sense of chemistry with the things we love. This chemistry may feel mysterious, but it boils down to the question of whether we enjoy the time we spend with the love object. Chemistry is driven by what psychologists call *intrinsic rewards*, which are the good feelings (e.g., pleasure, fun, and self-confidence) we experience while interacting with a person, using an object, or engaging in an activity. For example, the good feelings you get while singing are the intrinsic rewards of that activity. The word *intrinsic* comes from the Latin word for "inside of." If you enjoy singing, the pleasure you experience feels like it is inside of, or inherently part of, the activity. By contrast, the word *extrinsic* comes from the Latin for "outside of," so extrinsic rewards are motivations — such as being paid to do something — that feel outside of, or separate from, the activity itself. The two most common extrinsic rewards are money and social approval: that is, we often do things we don't enjoy because we're being paid to do them or because we want to please or impress someone else.

Intrinsic rewards are especially important in love. One man I interviewed loved his Prince tennis racket but only liked his Toshiba laptop.[12] Why? The tennis racket was intrinsically rewarding; he enjoyed the experience of playing tennis with it. But the laptop was extrinsically rewarding. He liked the results of using the laptop but didn't enjoy the experience very much.

Most things that people love provide a combination of intrinsic and extrinsic rewards — for example, cars are fun to drive (an intrinsic reward) and get us to work on time (an extrinsic reward). Yet when people talk to me about why they love something, they spontaneously mention intrinsic rewards more than 80 percent of the time — and only mention extrinsic rewards less than 10 percent of the time.[13] People do this, I believe, because even though the love object may provide some extrinsic rewards, people don't see these rewards as highly relevant to the question of why they love something.

One reason why people don't see extrinsic rewards as strongly related to love has to do with the familiar question "Do you love me or are you just using me?" For example, a woman once explained to me why she liked, but did not love, her workout shoes.[14] She thought they were high in quality, but she didn't enjoy the experience of exercising with them (she didn't find them intrinsically rewarding). She realized that what she really loved was looking fit — that's what she found rewarding. The shoes were just a means to an end — a tool to help her get fit, but not the thing she truly cared about, not the thing she loved. So with regard to that familiar question about loving versus "just using" something: if we *only* get extrinsic rewards from something, we feel we are just using it.

Intrinsic rewards also have a lot to do with why people do or do not love their jobs. In my interviews, the people who tell me they enjoy their work and find it deeply meaningful (it gives them intrinsic rewards) usually say they love their jobs. Even if the job is poorly paid, they often say things like "I love my work, but I sure wish it paid better."[15] Whereas I've never had anyone who is well paid but doesn't enjoy her work tell me she loves her job.

You might think, then, that extrinsic rewards such as salary don't generate love, but it's more complicated than that. What's interesting is the way that intrinsic rewards can "validate" extrinsic rewards. It's as if there's a checkbox next to INTRINSIC REWARDS on a form, and once that box is checked, extrinsic rewards appear as additional valid reasons to love something. The hypothetical person I mentioned above who said "I love my work, but I sure wish it paid better" would love her job even more if she were also happy about the salary. Extrinsic rewards can't *substitute* for intrinsic rewards, but extrinsic rewards can *add to* intrinsic rewards.

To cite another example, in the movie *Crazy Rich Asians*, the heroine falls in love with a guy and later learns he's fabulously wealthy. Lucky her. The money is an extrinsic reward of the relationship. If she'd known he was rich the whole time, she might have wondered, "Do I really love him? Or am I just using him for his money?" But because she didn't know he was rich when they fell in love, she is confident that her love stems from the way he makes her feel (an intrinsic reward). Learning that he is rich just makes the whole thing even better. Getting strong intrinsic rewards from something, in a sense, validates love. Then, if the love object also provides you with extrinsic rewards, that makes your love even stronger.

INTRINSIC REWARDS AND YOUR AUTHENTIC SELF

There is a second reason that intrinsic rewards are so important in love. Just as experiencing chemistry with a person on a date suggests that you two are a good fit, feeling pleasure when engaging with an object suggests that it is a good fit with who you are as a person, an authentic part of your self.

In studying romantic love, psychotherapist Nathaniel Branden[16] concluded that "psychological visibility," or the capacity "to see ourselves in and through the responses of another person," is a basic motivator of love between two people. Psychological visibility can also motivate our love for things, such as this woman's love of writing.[17]

> It's something that I feel constantly surprised and kind of tickled and delighted [over] — when I find the words and the expressions and the ideas I never knew I had and I certainly can't express verbally. I see it coming out on the page. I guess the reason I love it is . . . it tickles a place inside me I don't think I knew I had before.

As this woman writes, words and ideas she never knew she had emerge onto the page. This activity she loves isn't just a process of self-expression; it's also a process of self-discovery.

An interview I conducted with a man I'll call Joe provides a good example of how this process of self-discovery is closely linked to the distinction between intrinsic and extrinsic rewards. Joe is a professional musician who played in a rock band that included choreographed dancing as part of its performance. Most people I interview who raise the topic of dancing really love it. But Joe was an exception. Here he talks about why he doesn't love performative dancing.[18]

> JOE: The reason I started [dancing] is for the same reason I did music and comedy and everything else — I wanted to impress people, [make them] really like me and give me all this adulation and love. And you can get some of that: people just come up and go, "That was great, that was great." But one of the things I learned by doing this [is that] it doesn't really fill up any void. It's kind of like when people tell you they like your dancing . . . my reaction is sort of like, that's not me.
>
> ME: Dancing wasn't "you" — what do you mean by that?
>
> JOE: It's sort of like when somebody likes you for your car — that's an analogy I came up with. It's like if a girl went out with you and said, "I really like your hip car." Well, it's like, "Fuck you! Get out of here!" You know, it's got nothing to do with my real personality.

Because Joe's dancing was extrinsically motivated (that is, motivated by a desire to impress other people, not by enjoyment of it), even when he got the kind of positive response he wanted, he felt that the appreciation wasn't really about his authentic self.

"ATOMIC LOVE": HAVE YOUR CAKE AND EAT IT, TOO

I realize that the phrase "atomic love" sounds like the title of a terrible romance novel or perhaps an equally terrible 1970s disco song.* But it is an apt metaphor for a powerful phenomenon that can create passion in the love of things.

People today both get to and have to decide who they want to be. Deciding who we want to be is made more difficult when we are attracted to two or more identities that we see as conflicting with each other. For example, consider these comments from Cathy Guisewite,[19] creator of the *Cathy* comic strip:

> The best gift I ever received was a two-part gift that I got for Christmas when I was ten years old — a bride doll and an electric train. They were not things I'd ever told anybody I wanted. But they were like my secret-heart desires because those exact opposites were who I was and what I loved.

---

* It turns out that *Atomic Love* is also the title of a 2003 animated short film that is "all about cosmic love as woman and machine share space fries," according to IMDb.

The electric train was who I appeared to be. I loved everything mechanical; I wished I was a boy. . . . The bride doll was my more romantic side that I think wasn't as apparent, but it became my time with mom, making doll clothes.

Those two sides of myself are, for better or for worse, the "psychotic" conflict inside that has fueled almost all of my creative work. And if you look at the history of *Cathy*, it really was written out of that conflict in the beginning.

Here Ms. Guisewite describes a typical identity conflict. In her there were two "exact opposites" that "were who I was and what I loved." She experienced this as an identity conflict because, in her mind, the tomboy identity and the feminine identity were incompatible with each other.

There are several unsatisfying ways to resolve these identity conflicts. One bad option is to choose one identity and banish the other: being only the tomboy or only the feminine girl. A second bad option is to look for a compromise partway between the two identities. For example, Ms. Guisewite could have developed a gender-neutral persona that was halfway between tomboy and feminine girl. But in this type of mushy compromise, people often lose the good things they like about the conflicting identities. A better but still less-than-ideal option is to go with the "I contain multitudes" approach. In this approach, you accept that you contain many conflicting identities that don't really make sense together and just live with it. Many of us prefer this to either giving up one identity or creating a mushy compromise. But this is only a partial solution, because people like their identities to feel coherent and make sense to them.[20]

What people really want is to keep both identities but find a way of thinking about them so they no longer feel in conflict with each other. For example, what if Ms. Guisewite found a way to be both 100 percent tomboy and 100 percent feminine girl without experiencing a conflict? That is to say, instead of an either-or decision, people want a both-and option that lets them have their cake and eat it, too. The reason Ms. Guisewite loved the gift of the train and the doll was that they helped her move toward *both* a tomboy *and* a feminine-girl identity. In her case, it took two different gifts, one for each of her identities, to do this. But what if she had found a single love object that showed her a new way of thinking about these two identities in which they no longer seemed to conflict with each other? Now, that would be a neat trick.

This does happen, and when it does, I call it "atomic love." In atomic reactions, tensions between various parts of an atom are released, creating enormous amounts of power. In atomic love, tensions between conflicting parts of a person's identity are reconciled, generating enormous emotional power and creating particularly strong love.

One woman I interviewed, whom I'll call Pam, was raised to value a kind of upper-class, old-school femininity. She grew up near Hyde Park, London, and went to school "with the diplomats' children." As she recalled, this brought her into high society.[21]

> It was not unusual for [my parents] to get invited
> to the fall season — to, you know, the deb balls. . . .

> My parents used to take me out with them in the
> evening. . . . They didn't believe in babysitters at
> that time.

Her parents reinforced that high-society identity and gave her gifts that fit that lifestyle. Although she internalized a possible identity for herself as the kind of person who went to London debutante balls, her life eventually moved quite far from that. When I interviewed her, she was in her twenties, living in a cheap yet hip neighborhood in Chicago, trying to make it as a composer of film scores. Her main identity might have been described as progressive, bohemian, artsy, and intellectual. Yet she didn't want to totally let go of the upper-class socialite image that lived in her memory. She wanted some way to bring that past life together with her present.

Her beloved collection of Lucite handbags from the 1950s and '60s helped her do this. I had never heard of such bags before meeting her. But when Lucite was new, it was considered a high-end material and was used to make luxury handbags. She had received the first bag in her collection as a gift from her mother. Said Pam, "She gave me a very beautiful bag. It was the bag that she bought to go on the first date with my dad, a beautiful black lacquered Chanel bag." Later, she bought additional bags in this style. These bags were also a perfect fit for her artsy bohemian identity. They were elegant, very "off the beaten track," and bought cheaply at rummage sales. Remarkably, these bags were able to perfectly represent both the high-society femininity of her childhood and the artsy, bohemian feminism of her later life without compromising either. No wonder she loved these bags.

Another woman I interviewed, Cindy, had grown up on a ranch in Nebraska.[22] But she, too, had found her way to Chicago, where she became a successful executive. Cindy experienced an identity conflict between her rural rancher self, who had spent her summers on a tractor, and her urban sophisticate self, who lived in a downtown high-rise. She deeply loved her family heirloom rustic antiques, which she had brought with her from the ranch.* To her good fortune, this style of antique was also very much in fashion among her urban friends, many of whom owned similar furniture even though they didn't have a personal connection to it. Part of what made Cindy love this furniture was that it allowed her to maintain her rancher identity and enhance her urban sophisticate identity at the same time, without compromising either.

One last example: consider the popularity of Rolex and similar watches among businessmen in Singapore and around the Pacific Rim. These businessmen face an identity conflict. On the one hand, most of them are of Chinese descent, and traditional Chinese Confucian values say that a good person is hardworking, patient, self-sacrificing, and a wise investor for the future. Many Chinese businessmen strive to create an identity for themselves that reflects those values. On the other hand, research[23] has shown that in a collectivist culture, gaining honor by showing one's success is even more intensely valued than it is in the West. And one's success is most convincingly established by wearing expensive status symbols. This creates a tension between the frugal investor identity and the visibly rich person identity. What is a successful businessman to do?

---

* See pages 35–36, where Cindy is quoted talking about this furniture.

Rolex and similar watches costing $10,000 and up (way up) are immensely popular in Singapore, which is typical of East and Southeast Asia, making these regions by far the largest markets for luxury timepieces. That is partly because these luxury items offer people a have-your-cake-and-eat-it-too solution to this identity conflict. When I spoke to people about their Rolex, they always told me what a good investment it was.[24] "Yes, it was a big purchase," they would say, "but it has already gone up in value." In Singapore, there was a widely shared belief that these watches were dependable investments. One man, who stood out from the crowd by *not* wearing a Rolex, said that people frequently try to persuade him to buy one, stressing what a good investment it is. This belief that a Rolex is a good investment allows people to reconcile the two otherwise conflicting aspects of their identities. In their view, buying a $10,000 watch doesn't make them profligate hedonists who lavish luxuries on themselves. Quite the opposite; they are shrewd investors making good long-term decisions. And yet at the same time, they are able to display their financial success and earn the honor that comes with it.

ADS AND IDENTITY

We see ads all the time that don't seem to convey much information about the products they're promoting. Advertising for high-end perfume is a good example. You might reasonably think that the key piece of information in a perfume ad would be what it smells like. But most ads for expensive perfumes don't even try to describe the smell. They just show a beautiful model with a highly stylized look, often in a passionate embrace with another stunning model, or perhaps walking down a Paris street or standing in a rain forest. What's going on?

In classroom discussions, it's common for students to say that the existence of these ads shows that the world is full of idiots. As these students see things, the ads work by telling people, "If you buy the product, you will magically come to look like the model." And only an idiot would believe that is going to happen.* Yet perfume companies keep making these ads. So the ads must work on lots of people. Therefore, the world must be full of idiots, or at least highly gullible consumers.

The problem with these students' argument is that it starts with a faulty premise. For these ads to boost sales, they don't need to convince consumers that the products come with a magic beautification wand. Despite all appearances to the contrary, the ads aren't about physically transforming the consumer. What are they about, then?

The brilliant fashion designer Tom Ford was asked by Lynn Hirschberg of the *New York Times*[25] what went through his mind just before his fashion shows. In Hirschberg's words, he wanted "the fashion press and the buyers and, finally, the customers [to] look at these clothes on these girls and think, I want that life." Note that he didn't hope they would want "that outfit"; he hoped they would want "that life." Earlier in this chapter, I said that buying a shirt isn't just about finding a garment that fits your body; it's also about finding something

---

* The truth of the matter is even more extreme. The photos in these ads involve so much makeup, careful lighting, and computer retouching that if you saw the models in real life, you probably wouldn't recognize them. So not only don't the products make *you* look like the people in the ads, they also don't make the *models* look like the people in the ads. It's largely photo editing that does that.

that fits your identity. Well, the purpose of these perfume ads, and countless similar ads for other products, is to get consumers to make a mental link between the brand and a certain kind of person who leads a certain kind of life and therefore has a particular identity. The models in the ads are intended to represent that identity. If you spelled out the implicit message in these ads in plain language, it would sound something like this.

> Hey there, consumer. Take a look at these beautiful people. Imagine who they are. Imagine the lives they lead. Our brand can help you create and express that identity. No, we can't promise that you'll be that good-looking. And we can't promise that you'll have that much money. But you still can be, or perhaps already are, your own version of that type of person. And if so, our brand is "you."

For these ads to sell a product, all that needs to happen is for consumers to look at them and make a mental connection between the brand and the kind of person depicted therein. If consumers look at an ad and think, *They're my kind of people,* those consumers will come to see the brand as right for them.

This type of advertising is particularly common for expensive products. High-end brands aren't trying to get lots of people to buy cheap products from them; they are trying to get a few people to buy expensive products from them. And to do that, they need to offer potential customers something more than a good product, because there are lots of good products out there selling for less. They need to offer customers something they will *love.* And since one of the main drivers of loving something is seeing it as part of your identity, high-end brands use these ads to build the perception that their brands are part of an identity that their target customers are striving to attain.

I'VE SAID IN THIS CHAPTER that as our society became richer after the Industrial Revolution, people got more choice over how they wanted to define their identities. Choosing and then creating an identity entails work; it has become the project of the self. At some points in our lives, such as young adulthood, we think about this project a lot. But even when this project isn't at the center of our attention, it still lurks in the background of everything we do.

I've also asserted that if we see something simply as a tool for gaining extrinsic rewards, we don't feel that we love it. In order to love something, we need to also see it as intrinsically rewarding, which means that we enjoy the process of interacting with it. One big reason why intrinsic enjoyment is so important is that we see it as a signal that the thing in question fits with our internal authentic selves.

Since our enjoyment of certain things plays such an important role in our love for them, this raises the question: Why are some things more enjoyable than others?

7

Enjoyment and Flow

De gustibus non est disputandum
(There's no accounting for taste).

— LATIN PROVERB

PEOPLE WHO COOK TRADITIONAL CHINESE FOOD USUALLY EN-deavor to use the whole animal — including, sometimes, its eyes and genitals — in dishes that feature meat and fish ("Waste not, want not," as my mom used to say). Some Chinese dishes also incorporate animals, such as scorpions, that are not part of the typical Western diet. In the years when travel between the United States and China was relatively uncommon, I heard an interview with a group of China's most renowned chefs who were touring the United States, dining at many of the country's best restaurants. For most of them, this was their first serious taste of Western cuisine. What I found most memorable was how disgusted they were with the idea of eating rare steak. Thinking about the beef dishes I've been served in American Chinese restaurants as well as in China, I realized that any beef in them is usually cut into small pieces and cooked through. I had always found some foods from other cultures unappetizing but was naively surprised to find that non-Westerners find some of my favorite foods to be just as gross. This is but one example of a much larger phenomenon. Our own tastes usually seem so obviously enjoyable to us that we don't wonder about why we like them, whereas other people's tastes can seem so strange that they defy all explanation. In either case, we have little understanding of why we, and other people, enjoy certain things but not others.

Hence it is often said, "There's no accounting for taste." In other words, the reasons why one person prefers yellow and another prefers blue are so random and enigmatic that it's futile to try to understand them. But I disagree. Accounting for taste is what I, and many other social scientists, do for a living. In this and the following chapters, I'll sketch out the scientific reasons why certain preferences are pretty universal while others differ so much between people. I'll also discuss the aspects of an object or activity that make it enjoyable to some people and unpleasant for others.

TASTES THROUGHOUT YOUR LIFE CYCLE

The way our tastes are formed and then change over our lifetime follows a predictable pattern.

Infancy and Childhood: Simple Pleasures

Cats can't taste sweetness: offer them a bowl of sugar, and they will walk away indignantly. But cats are wired to taste fat: give them sardines packed in oil, and they'll love you forever. Okay, maybe not forever, but they'll love you until they're done eating. You get my point.

Like our feline friends, we're also wired to enjoy many things that help us survive and reproduce. For us, these include sweet foods, sex, a good night's sleep, and walking into a warm room after being out in the cold. I'll call these "simple pleasures." Because these simple pleasures are genetically programmed, we enjoy most of them from the day we are born.

To understand how people relate to simple pleasures (say, a chocolate bar or a massage), let's start by thinking back to the Love of Things Quiz (page 10), which has thirteen questions, each addressing a different aspect of love. In order for your feelings for an object to qualify as "true love," you need to give the object a high score on almost all thirteen questions. Simple pleasures normally score high on the "I find this enjoyable" question. However, from there, our relationships with simple pleasures usually follow one of two patterns.

In the first pattern, people enjoy the simple pleasure, but the potential love object gets low scores on questions such as "My involvement with this says something true and deep about who I am as a person" and "This does something that makes my life more meaningful." Even though people find these simple things pleasurable, they see them as too shallow or superficial to qualify as love.

In the second pattern, the things that produce simple pleasures do qualify as love. In these cases, the love objects score high on both the "I find this enjoyable" question and the questions about identity and meaning. For example, a person might love Girl Scout cookies partly for the simple pleasure of the flavor and partly because she has lots of fond memories of her own experiences as a Girl Scout. These mental associations make the cookies deeply meaningful and therefore worthy of love.

Another simple pleasure is the joy we experience when we achieve a goal or otherwise get something we want. We can see this even in very young children — for example, in the elation they feel when they finally learn to walk. A friend of mine, Scott Foster, had the opportunity to speak with the film director Krzysztof Wierzbicki,* who shared with him the secret to making a movie people will love: get the audience to want something really, really badly . . . *and then give it to them.* For example, the first 90 percent of an action film ramps up your desire for the hero to defeat the villain. Similarly, in romances, the audience is led to want the two lead characters to overcome the obstacles to their relationship and recognize their love ("Why can't they see how perfect they are for each other?"). In both these genres, just before the movie ends, the audience finally gets what it wants. In these films, the endings are usually predictable. Much of the director's skill, then, lies in finding creative ways to pump up the audience's desire for the inevitable climactic scene.

Most of the things that give us pleasure — our favorite movies, music, hobbies, sports, and so on — are things that differ among individuals and cultures. By contrast, simple pleasures — such as walking out of a freezing winter day into a warm room — are unusual because in their most basic form, they are natural, innate, and culturally universal. Other things that give us pleasure, such as foods that appeal to us beyond their being sweet and fatty, can seem so obviously enjoyable that we assume

---

* In this conversation, my friend also mentioned my view that virtually any film would be improved by the addition of space aliens. To my pleasant surprise, this did not end their conversation right there and then.

they are natural and innate (which leaves us wondering what's wrong with people who don't like them). But in fact, our enjoyment of these things isn't innate; it only seems that way because we learned to like them at a young age.

For example, a chef friend of mine, Amelia Rappaport, told me a story about a man who came into a fancy Boston hotel to arrange a big wedding reception for his daughter. The hotel was well known for its chocolate cake, and the customer tried some to see if it would make a good wedding cake. The customer thought the cake was good but somehow not quite right. So the chef invited him to come back to try other recipes. Unfortunately, the customer had the same "not quite right" reaction. The chef started to worry that she might never find the right recipe. Later, while grocery shopping, the chef had a flash of insight: the customer had grown up eating box-mix cakes. So without telling the customer what it was, the chef prepared a cake from a Duncan Hines cake mix. "Perfect!" the customer declared. That's exactly what he had been wanting.

In this story it's the father, an average guy rather than a food expert, whose tastes were formed by eating box-mix cakes as a child. But this tendency to like the things we grew up with can be present in experts as well. I remember a news story about expert food critics who arranged a blind taste test of dozens of different brands of ketchup. The winner was America's best-selling brand: Heinz. As one judge said, "It just tasted the most like ketchup." Of course it did! These food critics grew up on Heinz ketchup. So as adults participating in a blind taste test, they found that it tasted "right."

This process of forming lifelong tastes in childhood goes all the way back to our earliest experiences. For example, the food a woman eats affects the flavor of her breast milk. Biopsychologist Julie Mennella and her colleagues[1] found that babies have a tendency to accept solid foods, such as carrots, if their mothers have eaten those foods just before breastfeeding them. (It gives the phrase "like Mama used to make" a whole new meaning.) Mennella's research has taken this principle one step further, finding that if women drink carrot juice during pregnancy, the fetus experiences the carrot flavor in the womb. Even if the women stop drinking carrot juice after giving birth, their babies will still be receptive to carrot flavors when they start eating solid foods months later.

Early taste imprinting isn't limited to food. For example, many children learn to like their parents' favorite musical styles because they hear them so often. For some people, these early musical tastes stay with them throughout their lives, but for other people, not so much. This is because, when it comes to what we enjoy as adults, these early experiences are the beginning, not the end, of the story.

Adolescence Through Early Adulthood: Acquiring New Tastes

If our tastes were completely determined by our genes and early childhood experiences, we'd spend our lives eating chicken nuggets while watching *Teletubbies*. Fortunately, as we grow, our tastes grow with us.

As teenagers, we start increasing the amount of time we spend with our peers and decreasing the amount of time we spend with our families. During adolescence, our expanding tastes are powered in part by sheer opportunity as we get exposed to all sorts of new music, movies, foods, games, and so on, and as we discover new things to love.

While we are trying these new things, we are also working to define our identities in ways that distinguish us from our families. To do this, we focus on the things we have some control over, such as music, clothing, video games, hairstyles, and cell phones, and use them to signal which social group we belong to (or want to belong to). During these years, a lot of this exploration takes place within a group of friends. Discovering new things together strengthens bonds within the group. Neuroscientist David Rosen,[2] who studies music preferences, has suggested that this process may help account for the particularly strong bonds that people often develop with their friends at this age.

In fact, this exploratory phase has a lot to do with how I came to write this book. When I was in high school, I had some musically precocious friends who played jazz professionally in local clubs. We'd hang out and listen to this music. In the beginning, jazz sounded to me like meaningless musical squiggles. Eventually, though, it stopped sounding like random collections of notes and gradually started sounding like . . . well, music. I became fascinated by the process of how we learn to like things, and I continued to experiment with the process by consciously teaching myself to like all sorts of things, including cars, classical music, country music, art, football, and a plethora of food and drink. With each new thing, I made a point of paying attention to how and why my tastes changed over time.

One of my first experiments was teaching myself to like cars. I had never been interested in cars, but I figured that since millions of people really love them, maybe I was missing something. I started by noticing what they looked like and deciding if I liked what I saw. It didn't take long before I began to enjoy this process, and I formed detailed opinions about which cars looked good and why. But to understand cars, you can't just look at them; you need to drive them. I started visiting auto dealerships, pretending that my dad was planning to buy me a car and test-driving the sporty models. I started with the affordable brands and worked my way up to BMW and Porsche.* As I got older, I decided that lying to salespeople about my intentions was not okay, but I also discovered that it had become totally unnecessary. If I went into the dealership when it wasn't too busy and told the salesperson, "I have no intention of buying a car, but I am curious about the such-and-such model and would like to drive one," he or she would photocopy my license and hand me the keys. I love cars to this day (which helps explain all the car-related examples in this book). Eventually, my interest in why people enjoy certain things, and how we develop our tastes, led to the research out of which this book was born.

Middle Age and Later: Coasting on Autopilot

As people hit their twenties, they often cultivate grown-up tastes. For example, in my opinion, nothing goes better with a piece of chocolate cake than a glass of cold milk. But if I'm in a fancy restaurant, I'll probably choose coffee. Milk is literally the earliest childhood taste. It just doesn't seem appropriate in a sophisticated adult setting.

---

* In retrospect, I'm sure my ability to test-drive expensive cars as a college student had a lot to do with being a white kid at a university that attracted plenty of students whose parents actually did buy them fancy cars.

The exploratory phase that starts in the teenage years may last into a person's thirties or forties. But our openness to new things slows down dramatically as we get close to middle age. Neuroscientist Robert Sapolsky[3] has shown that our tendency to stop developing tastes for new things is linked to physical changes in the brain that occur around this age. The introduction of sushi into American cuisine in the late 1980s provided Sapolsky with excellent circumstances in which to test his theory. He discovered that people who were thirty-nine years of age or older when sushi was introduced rarely developed a taste for it, while those who were thirty-eight or younger often grew into ardent sushi lovers.

Have you ever wondered why so many people are convinced that whatever music happened to be popular when they were young is the best music ever created? Sapolsky found that most people were younger than twenty when they first heard the type of music they would love for the rest of their lives. Moreover, if people are thirty-five or older when they first hear a new style of music, there is more than a 95 percent chance that they won't like it ("That's not music; that's just noise!"). This was confirmed in a study by the music-streaming site Deezer, which found that the peak age for discovering new music was twenty-four and that most people stopped discovering new music around the age of thirty.

Older folks often believe that the reason they dislike new music is that it simply isn't very good. While the music of your youth was, undoubtedly, just as great as you think it was, we live in a world bursting with far more great new stuff of all kinds than in any previous era. I'll use music as an example, but the basic idea holds true for almost all forms of art and entertainment.

Both nature and nurture play a role in the making of a musical genius.[4] On the nature side of things, there is a genetic component to musical ability, and a few people are blessed with astonishing levels of talent. The average level of innate musical ability in the population may be increasing slightly. The argument for this claim is that musical talent is positively correlated with a person's overall IQ, and average IQ has been slowly increasing since it started being measured more than a hundred years ago, a phenomenon known as the Flynn effect.[5] But to be conservative, instead of saying that musical intelligence is increasing, we can assume that it has remained more or less the same. Since the population of the planet has been growing, and the percentage of the population with the genetic ability to be a musical genius has remained about the same, it is reasonable to conclude that the number of people with extraordinary raw musical talent has also been growing.

But becoming a musical genius requires more than genetics (nature); it also requires training (nurture). Musical geniuses of the past, such as Bach and Mozart, didn't just have natural talent. They also grew up in musical families that gave them an intensive musical education. University of Colorado geneticists Hilary Coon and Gregory Carey studied musical ability in twins and determined that while a person's genetics and the environment in which he or she were raised both influenced musical ability, the environment mattered more.

Today's environment is much more conducive to producing musical geniuses than any other social environment in human history. For example, getting a strong musical education is very important for nurturing a person's innate musical talent, and today there is more widespread formal musical education than there was at any time in the past. And this education is available to women as well as men.

But even great genetics combined with a top-notch education aren't enough. Musicians and composers still need to get their work published — either as sheet music or as a sound recording — and have it widely distributed. Today's large global music industry, aided by government and nonprofit funding, records and distributes the music of far more people around the world than it did in the past. Better still, it is now possible for people to make professional-quality audio recordings themselves. This point was fully brought home when Billie Eilish won five Grammy Awards for an album her brother had produced on his laptop in his bedroom.

When one considers the growing number of people with genius-level musical talent and an environment that is more nurturing of this talent than in the past, one can see why there is no shortage of great new music. But because our musical tastes tend to be set as young adults, and music styles are always changing, for people who are middle-aged or older it sure can feel like the quality of music is decreasing as time goes on.

Continuing to find new things to love as you age makes your life a little richer and more exciting. Furthermore, learning new things throughout your lifetime reduces your chances of developing Alzheimer's disease and dementia. But acquiring new tastes past middle age does require a bit of a "manual override" to counteract the tendency to coast along with the same old same old. Specifically, when you try something new and don't like it, you need to keep trying it until it starts to grow on you. We do this all the time as parents when we coax our kids to eat unfamiliar foods. Young adults also do this as they expand their tastes, but it frequently happens naturally in their social settings, so it isn't something they necessarily need to consciously work at.

The need to try something several times in order to develop a taste for it is part of a common pattern in the way people respond to new things. Consider clothing, for example. When people first see a new style, it often looks weird, ugly, and a bit embarrassing to wear. But later, after they've seen it several times, they start to like it and may even buy it. Then, as time goes on, they change their minds again, deciding that they no longer like that style and wondering what in the hell they were thinking when they bought it.

When this happens, we call it the "fashion cycle." The fashion cycle is influenced by advertising, but it isn't created by advertising. Nor is it limited to clothing. Even though no advertisement encouraged parents to name their babies Jaxson, for example, that name increased dramatically in popularity from 1999, when it was given to around eight children per year, to 2009, when it was given to more than two thousand children per year.[6] Over the same period of time, the girl's name Misty went out of fashion and dropped in popularity: it had been given to around 1,700 babies per year but ended up being given to just fourteen per year.[7]

So if fashion cycles occur without marketing, how do they originate? To answer that question, it's helpful to understand the nature of flow and repeated experiences.*

FUN, FLOW, AND THE FEELING OF LOVE

I explained in chapter 3 that though it is common to feel oxytocin-based affection for another person, it is less common to feel that same kind of affection for things. Yet loving things is

---

* The theory I explain in the following section is just one of several causes of the fashion cycle.

still a very emotional experience, and often these feelings include intense enjoyment and full immersion in a pleasurable activity. The psychologist Mihaly Csikszentmihalyi* researched life's most engaging and enjoyable experiences. He called these peak experiences "the flow state," so his theory is called "flow theory." The flow state has also been called "peak experience" and "being in the zone." These terms make it seem like flow is a rare experience, and at its most intense levels, it is. But at moderate levels, we experience flow quite regularly; it is just called "having fun."

Flow is only one of several reasons why we enjoy things. But it's extremely important for understanding our love of things. Csikszentmihalyi's basic theory of flow has spawned a small army of researchers who have expanded it and refined it in various ways. As I have applied the theory to the things we love, I've also elaborated on it a bit. Below is my take.

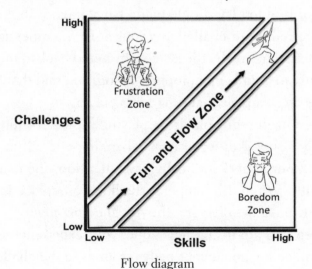

Flow diagram

---

* As I once heard him explain, his last name can be pronounced "Chick-sent-me-high."

To understand the flow diagram on page 171, imagine a pianist playing a piece of music. The vertical axis indicates how challenging or difficult the piece is. The horizontal axis indicates the pianist's skill level. If she is a beginning pianist (with low skills), and she tries to play something very hard (with high challenges), she will fail repeatedly and end up in the upper left corner of the diagram: the *frustration zone*. By contrast, if she is a fabulous pianist (with high skills), but the song is written for beginners (with low challenges), she'll be in the bottom right corner of the diagram: the *boredom zone*.

When an activity is too easy, you have lots of unused brainpower. All that extra brainpower starts generating extraneous thoughts and daydreams, so you feel distracted and disengaged from what you're doing. I've witnessed this in action while working at a restaurant that had a piano bar. The pianist's job called for him to play simple piano-bar standards, but the pianist preferred more challenging music. I remember saying to him, "Bart, that's really nice." He whispered back to me, "Is it? My hands are totally on autopilot. Before you said that to me, I wasn't even aware of what song I was playing."

Now imagine our pianist is playing a piece of music that is exactly at her skill level: it's challenging enough to require concentration, but if she really pays attention, she can play it beautifully. When this happens, she'll be someplace in the diagonal stripe of the diagram: the *fun and flow zone*.

When you are in the flow zone, your experience goes beyond just feeling good; your body responds to the challenge by making you feel energized and wide-awake. Because success requires your total concentration, you don't have extraneous thoughts or worries floating around in your head. Your brain is so focused, in fact, that it even co-opts the brainpower it usually devotes to keeping track of time and applies it to the task at hand. Time seems to fly because you haven't been paying attention to it.

People sometimes have a misconception that when it comes to tasks, easier is better. But flow theory tells us that activities below your skill level are boring. Researchers Maria Rodas and Carlos Torelli of the University of Minnesota[8] demonstrated this through an experiment in which two groups of participants tasted a new kind of gummy candy. One group ate the candy with their hands, but the other group was told to eat the gummies with chopsticks. Eating the gummies with chopsticks made the experience a little more challenging, which made it more interesting and fun, which led the chopstick users to like the gummies more than the people who used their fingers.

In 2019, the global video-game industry earned more than twice as much as the movie and music industries combined.[9] The reason for this level of success is that video games are flow machines. They present you with a series of challenges: destroy the monster, line up tiles of the same color, arrange letters into words, and so on. The secret of good game design is to keep a player's brain in the flow zone, where the challenge is hard enough to avoid being boring but not so hard that it's frustrating. The big advantage of video games is that as people improve, the games get correspondingly harder, so the challenge can always stay at a fun level. To get this right, video-game companies study people as they play, making sure that the increases in difficulty correctly balance a player's increasing skills. The gaming company Valve Corporation[10] is taking this to the next level by having people play their games while wearing neural sensors on their heads that can detect degrees of concentration and emotional responses. At the time of this writing, this information is being used to fine-tune the games so they will be enjoyable to a typical consumer. But since many of us aren't typical consumers, the company is also working on ways to use these neural sensors to customize the gaming experience as people are playing. In this scenario, players would wear a helmet containing the sensors, which would monitor the players' brain activity and adjust the level of difficulty to keep them in a flow state as they play.

The basic idea of matching a game's challenges with a player's skill level has been around for a long time. For example, there is a classic progression of card games that gets harder as it goes along. The common children's card game war is simple, yet it teaches the basic principle of taking tricks. When war gets dull, people can learn a more complex game called euchre, which is followed by spades and hearts, then whist, and finally bridge. You can also control the challenge level of a game by choosing an appropriate opponent or, in games such as golf, by giving good players a handicap. But none of these approaches comes close to the ability of video games to match a player's skill level with the optimal level of challenges.

Why Most Jobs Aren't as Fun as Video Games

People who love their jobs often spend a decent chunk of their working hours in a state of flow. Yet for too many of us, this is not the case. This is true despite the fact that jobs and video games both present us with challenges that we try to overcome using our skills. So why are so few jobs reliable sources of flow?

First, video games are extremely good at matching challenges to our individual skill levels. But at work we are often assigned tasks that are either frustratingly hard or tediously easy.

Second, there is an additional aspect of flow theory that is often overlooked: activities produce enjoyment if they include clear goals and offer rapid, understandable, and accurate feedback. In video games, your goals are clear, your score accurately reflects your performance, and you can see your score increase as soon as you do something good. But at our jobs, it's sometimes unclear what we need to do to succeed. Our performance may be inaccurately measured. And performance reviews may come as infrequently as once per year, if ever.

Third, we approach games with a more playful, "pro-fun" attitude than we usually adopt at work. If you think about it, most games involve doing utterly inane things, such as lining up three tiles of the same color. It would be easy to say, "That's stupid; I'm not going to do that." But instead, we make an implicit bargain with the game: we agree to care about the ridiculous task it gives us in exchange for the fun it provides. I suspect that for some of us, there are at least a few tasks at work and at home that we might enjoy if we approached them with the same attitude we take toward games — an attitude in which we seek out whatever fun there is in the task at hand.

Finding Flow in Unexpected Places

What I find so useful about flow theory is that it can help explain the enjoyment we get from a wide variety of things, including reading, eating, watching movies, looking at art, and listening to music. In order for you to enter a state of flow, the activity you're engaged in must present challenges, and you must use your skills to try to overcome them. Let's start by examining those challenges.

It's obvious that when people play piano or ski, they face challenges and that their success depends on their level of skill. But a lot of things we enjoy, such as eating, watching a movie, and listening to music, don't seem particularly challenging. And it isn't obvious that our enjoyment of them depends on our level of skill. Yet these activities are, in their own ways, challenging. And enjoying them does depend on how skilled a person is at meeting these challenges.

The core challenge involved in all these activities is comprehension, which occurs when your brain receives a bunch of sensory stimuli (e.g., smells, sounds, and patterns of light) and turns them into a coherent, meaningful experience of the world. When you're watching a movie, this not only means recognizing that the rectangular shape on the screen is a house, it also means knowing who the characters are and following the plot. When you're listening to music, comprehension means hearing the sounds *as music* rather than just a bunch of noises.

Your experience of the world is a mental model of the physical world around you. When your brain creates this model, it's a little like solving a jigsaw puzzle. All the sights, sounds, feelings, smells, and tastes that your brain receives from your senses are like individual puzzle pieces that your brain assembles into a coherent model of reality. Your brain continually checks the accuracy of that model by making little predictions and seeing if those predictions are accurate.

For example, as you walk down a flight of stairs, your brain predicts when your foot should touch the next step. So long as your foot reaches something solid at the right time, your brain knows that all is well, and your conscious mind isn't even aware that these predictions are being made. But have you ever walked down some steps without paying much attention to where you're going and underestimated how many steps there were? You put your foot forward, expecting it to hit the floor, but instead it just keeps going. At that moment, when your brain anticipates that your foot will touch the solid floor but it doesn't, your brain sounds the alarm, and a shock pulses through your body.

Or have you ever taken a sip of what you expected to be water but turned out to be something strongly flavored instead? The shock you feel is very different from what you would have experienced if you had known ahead of time what the beverage was. I still remember the time when, as a kid, I snuck a piece of chocolate from the kitchen counter where my mom was baking only to bite into it and discover that it was a horrible bitter substance called "baking chocolate." I still recall feeling a burst of alarmed shock. These feelings of shock illustrate that the brain is continually making unconscious predictions about what's about to happen, and this process only becomes conscious if something unexpected occurs.

Comprehending what's going on during typical daily activities, such as walking down the street or brushing your teeth, is not very challenging, which is why you may find those activities a bit boring. One basic function of entertainment (books, music, movies, and so on) is to provide your brain with a stimulus that is more challenging than your normal activities are. Because following a movie is at least moderately challenging, it requires your brain to be fairly active, which gets you out of the boredom zone and into the flow. Forms of entertainment differ from one another in how challenging they are to comprehend.

For example, a movie that follows a standard plot formula is not very challenging to follow, whereas a movie that breaks genre conventions is more challenging.

We're used to the idea that art experts are highly skilled at understanding Renaissance paintings, for example, whereas nonexperts are less skilled. But we don't usually think of other types of entertainment, such as TV shows, as requiring skill to enjoy. The fact that watching TV is a skilled activity was driven home to me one afternoon as my children, then ages two and four, watched *Teletubbies*. My boys were enthralled as the four characters chased one another in a circle until they all crashed together and fell down. The characters did this over and over (and over again). That was the whole episode. My boredom was relieved only by thinking about what a good example of flow theory this would make. Preschool kids' skills at watching TV are still pretty low. *Teletubbies* was designed to be ultrasimple, to match preschoolers' developmental level: its challenges match their skills, and the show puts them in the flow zone. Clearly it works for little kids, but its challenges are so low that it is also rather famous for putting adults in the boredom zone.

The tables were turned years later when I allowed my kids to join me for an episode of the original *Law & Order*. Reasonable people can disagree about whether *Law & Order* constitutes quality television. But there is no disputing the fact that it is quantity television: you get a cop show and a lawyer show, both crammed into forty-five minutes. *Law & Order* moves very fast by skipping all sorts of plot details and letting viewers fill in the blanks based on their knowledge of how similar shows usually work. This can make the show somewhat challenging to follow if you haven't seen similar shows before. At the time, my kids hadn't watched many cop shows, so they didn't have the background knowledge needed to understand what was going on. Since the show's challenges far exceeded their comprehension skills, this put them in the frustration zone. Unfortunately, they tried to reduce their frustration by barraging me with questions. The task of following the show while answering their questions was above my skill level, so I landed in the frustration zone, too.

Flow theory can also help us understand why the kinds of things we are in the mood for change from time to time. Our skill at appreciating TV, music, and other forms of entertainment can fluctuate. For example, being tired decreases one's skills, which is why at night, when you're tired, your taste in TV shows can turn from serious dramas that take a lot of concentration toward light shows that take less mental effort to enjoy. Alcohol also decreases your skills. You'll recall that euchre is an easy version of bridge, which is why it is sometimes called "kids' bridge." But given its popularity as a postdrinking late-night game on college campuses, it might be called "drunk bridge."

As another example, let's compare listening to music in a symphony hall to listening to music in a dance club. The symphony hall is set up to maximize an audience's skill at following challenging music: everyone sits facing the musicians, and there are no distractions such as people moving around and talking to their neighbors. By contrast, clubs lower our music-listening skills: we are distracted by talking, flirting, dancing, and checking other people out; late at night, our brains may be tired, even if they're also revved up on caffeine; and we may have made ourselves pleasantly stupid with alcohol or other drugs. All this combines to temporarily lower clubgoers' music-comprehension skills, which leads them to enjoy relatively simple and repetitive music. Simplicity and repetition don't make club music bad; they make it well suited for enjoying in clubs.

What Makes Things Challenging?

Suppose you want to modify an object or activity to make it more (or less) challenging. What exactly would you change? If you are trying to modify, say, a marathon, it's obvious that a hilly course is going to be more demanding than a flat course. But for lots of objects and activities — especially forms of entertainment, in which much of the challenge lies in comprehending them — the choices are not obvious. I've found that the challenges something presents are usually a result of four factors: complexity, subtlety, intensity of stimulation, and the extent to which specialized knowledge is required to enjoy it.

Complexity

One reason why *Teletubbies* is less challenging to understand than *Law & Order* is that *Teletubbies* is less complex. A show with many characters and several story lines is complex, like music in which many instruments play different parts at the same time. One of the things I love about good cheese, for example, is how complex it is. Usually, to get that many flavors into a dish, you'd need to combine a lot of ingredients. But a good cheese can contain lots of distinct flavors all on its own. The more complex something is, the harder it is for your brain to make sense out of it — that is, the more challenging it is to appreciate.

Because complexity makes things difficult to understand, marketers usually limit the complexity of their advertising so that even a person who isn't paying close attention can understand what's going on. At the same time, they don't want their ads to be so simple that they bore people. I have seen proprietary research conducted by consultants who studied the complexity of many TV ads. They considered factors that affect how challenging an ad is to understand, such as the number of scene changes, how bright the colors are, how prominent the music is, how clear the narrative is, and so on. They then considered the connections between each of these elements and whether consumers understood and remembered the ad. This resulted in a "recipe" for an ad that's complex enough to be interesting but not so complex as to be confusing.

Subtlety

Along with complexity, things are made more challenging if appreciating them requires making subtle distinctions. For example, a novel that tells you what each character is feeling (e.g., "Jane was angry about having to pick up Sarah's kids") is less challenging than one that expects the reader to infer the character's emotions from the way the character is speaking ("It's always nice to see your kids," Jane said, "but a little warning would be helpful").

Cheese is not only a good example of complexity, it's also a great example of subtle distinctions. The good people at Cheese Science Toolkit have identified forty-seven common flavors in cheese. As you first learn about cheese tasting, you might classify flavors in general categories such as fruity, floral, yeasty, and "barnyard" — this is cheese, after all. Then, as your expertise increases, you notice subtler distinctions. Instead of identifying something as floral, for example, you might distinguish between various flowers. To use a less appetizing example, you'll start to notice the difference between subtle background flavors, such as horse blanket, sweaty sock, and cat pee (none of which, rest assured, has anything to do with actual horse blankets, sweaty socks, or cats).

Intensity of Stimulation

Some things, such as habanero peppers, horror movies, and superfast roller coasters, provide a particularly intense level of stimulation. These experiences are usually too extreme for small children. But over time, as we gain maturity, we tend to enjoy strong flavors, fast roller coasters, and other heightened experiences.

Intensity of stimulation and subtlety of stimulation may seem like opposites, but they often go together. For example, meals at gourmet restaurants often include complex combinations of strong flavors and subtle flavors. That combination of subtlety and intensity makes a dish challenging to fully appreciate.

If something produces too little sensation, there's not much to enjoy, but if it produces too much sensation, it can be unpleasant or even painful. People differ in what they consider the just-right intensity level for a pleasurable experience, and these differences have a strong genetic component. People with a high genetic need for stimulation tend to be extroverts who like a lot of variety and crave new things, whereas people wired for less stimulation tend to be introverts who "know a good thing" when they've found it and don't feel as much need for variety. People with a high need for stimulation are also at a high risk of getting into trouble with drugs, and they are more likely than average to have ADHD.

There are times when people like what appear to be unusually intense experiences, yet those appearances can be deceiving. For example, on average, men like louder music than women do.[11] Many people assume, then, that men, more than women, enjoy the subjective experience of loud music. However, on average, men have less sensitive hearing than women do. So they might need to turn up the volume in order to get the same subjective experience as women get. This basic principle underlies a difference between my wife and me when it comes to food. I envy her because she has much more discerning taste buds than I do. My relatively insensitive taste buds lead me to want strongly flavored foods. But it's not because I want a more intense experience of flavor than she does; it just takes more strongly spiced foods for me to have the same experience.

Habituation also plays a role in how much intensity people want. If you're habituated to something, it simply means that you get used to it over time, so you start to need more of it to achieve the desired effect. You may have heard that people get habituated to addictive drugs, but people can get habituated to a wide variety of things. For example, people find the ringtones on their phones most noticeable when the phones are new; the ringtones gradually become less noticeable over time.[12] And people who wear perfume or cologne find that the scent seems to weaken over time, even though other people may find it quite strong.

I encountered an amazing (and somewhat tragic) example of habituation while I was in Las Vegas with a friend, David Obstfeld. David has an uncanny ability to strike up deep conversations with strangers. We were in an Uber when he started talking to the driver, who, we learned, had seen a lot of combat in one of America's recent wars. He had become habituated to the intensity of his combat experiences, which had left him so emotionally numb that the only way he could "feel anything" was to go skydiving *without a parachute*. He'd go up in a plane with an old combat buddy of his who was also a highly skilled skydiver. He'd jump out of the airplane without a parachute, and his friend would wait for a bit and then jump out after him. By controlling the angle of his body, the Uber driver could slow his descent. The friend would then need to catch up with him and rescue him before he hit the ground. He said he did this around once every six months "just to feel alive."

A more typical example is that the more sugar you eat, the duller your taste buds get, leading you to add more sugar to your food to get the desired effect. I gave up all added sweeteners for a month to give my taste buds a chance to regain their sensitivity. I discovered that fruit tasted far sweeter than it had before, while many packaged and restaurant foods tasted revoltingly sweet. As much as I enjoyed fully tasting the natural sweetness in foods, supersweet foods are so common that I found avoiding them too inconvenient to continue. It didn't take long for my taste buds to revert to their formerly dulled state.

Habituation to sweetness is literally killing us. Restaurants and packaged-food companies have found that if their food is just a tad sweeter than their competitors' food, customers will like it better. This has led to a gradual escalation in the amount of sugar added to foods as companies try to "outsweet" the competition. As a result, nutrition researcher Elyse Powell found, American adults consumed an average of 30 percent more sugar in 2010 than they did in 1977.[13] This is largely because sugar has worked its way into foods that we don't think of as desserts.

As you read the following examples, I encourage you to imagine yourself with a teaspoon in your hand, spooning sugar from a bag into each of these foods: a cup of canned baked beans contains around five teaspoons of sugar; a cup of granola contains around seven teaspoons of sugar; a cup of flavored low-fat yogurt can contain around twelve teaspoons of sugar. This issue made the news when the Irish government, which taxes cake at a higher rate than it taxes bread, ruled that Subway's bread contained so much sugar that it qualified as cake for the purpose of establishing the sales tax rate. When I first learned that, my response was "Really? I've tasted Subway bread, and it doesn't taste sweet to me." But upon reflection I realized that's exactly the problem.

All this sugar is doing very little for our enjoyment of food, because our taste buds have been desensitized. Meanwhile, all this added sugar is, according to Johns Hopkins cardiologist Chiadi E. Ndumele, "an important contributor to weight gain"[14] and, according to a University of Utah study, a leading driver of the rising rates of diabetes.[15] This competition between food producers to make their products slightly sweeter than the competition is killing some of us, injuring many of us, and increasing medical costs for all of us.

Requires Specialized Knowledge

In addition to complexity and subtlety, things can be challenging to enjoy if getting the most out of them requires specialized knowledge. Some years ago, I teased a baseball-loving friend of mine by calling baseball "land chess." To my surprise, he agreed, explaining that he loved baseball because of its chesslike strategy. Thinking about a team's strategic choices is a common source of pleasure for baseball fans. I once interviewed a woman who told me how she came to love baseball.[16] She had started dating a minor league player who took her to a Cubs game. As they sat in the stands, he explained to her that because there was a man on second base, and because the pitcher was right-handed and the next batter was left-handed, all the defensive players were going to shift positions by a few feet in various directions when the next batter approached the plate. She didn't believe him. But lo and behold, when the batter walked onto the field, the players all moved around just as her date had predicted. She realized that there was much more to the game than she had been aware of, and she became curious to learn more. As she gained more specialized baseball knowledge, she eventually reached the point where her skill at watching baseball was a match for the complexity of the game, and she could appreciate what was happening. Only then did she really fall in love with it. That's why people like me, who don't know much about the game, find it so boring. But as one's baseball knowledge grows, one eventually reaches the point of flow, and baseball becomes very engaging (or so I'm told).

Like baseball, wine also requires specialized knowledge to get the most pleasure out of it. I once went to my regular wine shop, where the owner excitedly recommended a $20 bottle of French red to me. It cost more than I usually spend, but he seemed so enthusiastic that I figured, "Why not?" The wine was fine, but I didn't see what the fuss was about. Next time I went into the store, I told him about my "meh" reaction to the wine. He explained that it was from a certain region in France where the wines have a special and unique flavor. Good wines from there usually cost $50 or more, but the wine I tried was a good example of one of these wines, and it only cost $20. If only I had known. To fully enjoy that wine, I needed to know about that region; that is, I needed to know the wine's backstory.

Some of you may be thinking, *If you need to be told about a wine ahead of time to enjoy it, that means it doesn't really taste good and you're just pretending to like it to seem sophisticated.* This argument is an example of aesthetic formalism, which is the belief that what makes something good is contained within the thing itself. For example, formalists argue that what makes a painting good is contained on the canvas, so information about the artist and the painting's historical context is irrelevant. I understand the appeal of this point of view because I used to agree with it.

I rejected aesthetic formalism, however, as I learned about what happens in real life as people enjoy things. For example, people who love sports tend to know a lot about the players, including their lives off the field. Formalists would see this as irrelevant gossip, because to a formalist, all that should matter is what happens in the game. But it's not irrelevant gossip: it's *relevant* gossip. Knowing the players as people gives sports fans a complex and nuanced understanding of what is happening on the field. It also enhances their person-thing-person (in this case, fan-team-player) connection with the team, which makes watching the game a meaningful experience. It would seem ideological and overly rigid to tell sports fans that they shouldn't learn about the players, even though doing so would increase their enjoyment of the game. Similarly, whenever serious art lovers tell me why they love particular paintings,* their explanations weave together

---

* This is different from nonexperts who tell me about the paintings in their living rooms, in which case they only talk about the artists if they have a personal connection to them.

what's happening on the canvas with what was happening in the artists' lives and in the society at the time the works were painted. If you take a purely formalist approach and ignore the story behind how something was created, you strip out a lot of what makes people love art. What's the fun in that?

Repetition, Skills, and Enjoyment: "Play It Again, and Again, Sam"?

I introduced the discussion of flow by talking about a typical fashion cycle, in which new styles become popular and then unpopular over time. The diagram below helps explain that phenomenon.

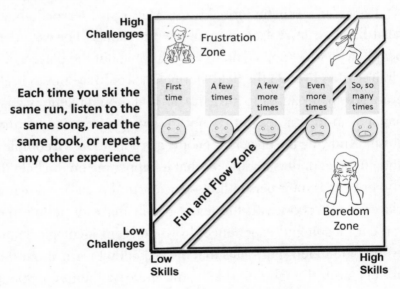

Flow and repeated experience diagram

Earlier in the chapter, I cited the example of a beginning pianist (with low skills) trying to play a very difficult piece of music (with high challenges), and I suggested that this would land her squarely in the frustration zone. There are two ways in which she might relieve her frustration: she could switch to a less challenging piece of music, or she could increase her skills. The diagram on page 188 shows what happens when our frustrated pianist combines both approaches. Instead of an expert-level Chopin étude, she switches to a less difficult piece, which is why the line of faces across the diagram is situated midway between low and high challenges. She also needs to increase her skills, which she will achieve by repeatedly playing the less difficult piece.

As you can see from the faces in the diagram, the first time she plays the new piece, it is still too challenging for her level of skill, and she remains in the frustration zone. But as she plays it more often, her skills improve, and she moves into the flow zone. Yet once she's mastered the piece, if she continues to play it repeatedly, her skills keep getting better, so she winds up in the boredom zone. That's how we get the basic pattern in the fashion cycle: at first, we like something more each time we experience it, but then it hits a peak, and after that we start liking it less with each repetition.

It's easy to see how a pianist's skill at playing a piece of music increases each time she rehearses it. But what is often missed is that the same thing happens each time we listen to a song or watch a movie. Just as the pianist's enjoyment of the new piece increases and then decreases as her skill improves with practice, our pleasure in listening to a piece of music also increases and then decreases as the brain's skill at comprehending the music improves with practice.

This same basic process also explains what happens when you hear a story in the form of a book, movie, TV show, play, or joke. However, most stories are written with the assumption that people will hear them only once. Therefore, the writers try to set the challenges at a level that will put most people in the flow zone the first time through. When we read a story, a lot of the challenge lies in predicting what the characters will do next. If you've already read a book once, the challenge is so drastically reduced on the second reading that you're unlikely to reread it. This isn't always true, though. One of the things that makes great literature great is that it has lots of subtlety and complexity. Even if you know how the story ends, there is still more to discover on repeated readings, so that can be enjoyable.

So far, I've been talking about the usual situations in which people get sick of things after a while. But what about people who have seen their favorite TV shows dozens of times? Why don't they get sick of them and move on? Flow is one source of the pleasure we get from things, but it is hardly the only one. We also enjoy things that trigger pleasant emotional memories. For example, here is a guy I interviewed talking about why he loves pizza.[17]

> Pizza is a biggie. Pizza has always been my favorite
> food, and ever since I was a little kid, I loved it. . . .
> When I was a kid, I remember eating my first pizza
> at the age of four and watching *King Kong*, and it
> was like this vivid memory, and from that point it
> was very important.

Old favorites remind us of past experiences and allow us to enjoy happy emotional memories. For some people, the pleasure of triggering these emotional memories is more than enough to compensate for the boredom of repeating the old favorite yet again.

When it comes to positive mental associations when listening to music, I've recently discovered a way to have my cake and eat it, too. I seek out covers of my favorite old songs, redone in innovative ways. For example, k.d. lang did a fabulous cover of one of my favorites, Neil Young's "After the Gold Rush." I fell in love with Young's original version in high school and have lots of positive memories associated with it. Many years later, k.d. lang's cover gives me the pleasure of triggering these positive associations, but because her arrangement of the song is so different from the original, I find it interesting to listen to repeatedly.

Positive emotional memories can be powerful enough to get us to love things that, without the memories, we'd probably consider unpleasant. For example, I recently heard a man explain that he loves the smell of fresh tar because it reminds him of the freshly tarred streets where he and his friends went Rollerblading: "Especially when the sun was really hot and just boiling that street, you would smell the tar."[18] Scent has an especially strong ability to trigger emotional memories because the part of the brain that processes scent is the limbic system, which also plays a big role in emotion. As Pamela Dalton, an olfactory scientist, explains, scent activates "the emotional memory system. . . . [So] memories that are triggered by odors . . . almost feel like you're being brought back in time."[19]

The fact that we like things that trigger positive emotional memories has an impact on our preferences, including our favorite colors. Research conducted by psychologists Stephen

Palmer and Karen Schloss[20] has found that our preference for various colors is largely based on how much we like things that our brains associate with those colors. For example, because the Michigan and Ohio State football teams are intense rivals, it's not surprising that Ohio State fans love the color red (their team color) and hate the color blue (Michigan's color) and that Michigan fans love blue and hate red. What is surprising, though, is how widely this principle applies in less obvious everyday situations, such as the discovery that people's love of the color yellow is linked to how much they like bananas.

THIS CHAPTER HAS EXAMINED FLOW, which works at an individual level, meaning that it works the same way regardless of whether people are alone or with others. In the following chapter, I'll continue to examine why people love the things they do but focus on the ways in which interactions between groups of people shape what we love.

8

What the Things We Love Say About Us

A people is an assemblage of reasonable beings bound together by a common agreement as to the objects of their love.

— SAINT AUGUSTINE, *THE CITY OF GOD* 19.24

THE THINGS ONE LOVES CAN PROVIDE CLUES TO THE NATURE OF one's character. For example, psychology professor Omri Gillath and his colleagues[1] studied the conclusions people draw based on the shoes other people are wearing. They found that people who wore pointy-toed shoes were correctly seen to be, on average, slightly less nice and accommodating (i.e., lower in the personality trait of "agreeableness") than people who didn't wear pointy-toed shoes. They also found that college students who were politically liberal were stereotyped as "hippies," and in keeping with that stereotype, liberal students were more likely to wear shoes that were seen as unattractive by other students (although I suspect those liberal students might have preferred to describe their shoes as comfortable).

Our shoes can even offer clues about our early childhood relationships with our parents. People who did not form strong childhood bonds with their parents can develop attachment anxiety, the nagging fear that others will reject them. This leads them to avoid shoes that look scuffed-up or damaged, lest they be viewed negatively by other people. This research calls to mind a woman I interviewed who said that it was unfair to judge other people by their clothing, but then added, "Except for their shoes, of course."[2] It didn't take much prodding for her to share her opinion — learned from her mother — that scuffed-up shoes were a sure sign of low moral character. I was glad that I was dressed "professionally" for the interview, without my usual footwear.

BIRDS OF A FEATHER...

The things we own or do tag us as belonging to what marketers call a "lifestyle group."[*] A lifestyle group is a much larger version of what, in high school, we used to call a clique — the jocks, the burnouts, the nerds, and so on. It's a subculture of people who tend to hang with others from the same group because they have similar tastes and opinions. Terms such as "yuppies" and "soccer moms" started out as marketing jargon for lifestyle groups. One popular system of lifestyle groups called Mosaic breaks the United States into seventy-one different segments, such as Young City Solos (young and middle-aged singles living active and energetic lifestyles in metropolitan areas) and Flourishing Families (affluent middle-aged families and

---

[*] Also known as a "lifestyle segment."

couples earning prosperous incomes and living very comfortable, active lifestyles).[3]

Even though most of us aren't familiar with the names given to these segments, we can usually recognize the people in them. This is particularly true for lifestyle groups that are associated with iconic products, such as people who drive Priuses and people who drive pickup trucks. Of course, many people who use these products don't fit the stereotype. But marketers spend hundreds of millions of dollars studying lifestyle groups because, statistically speaking, there are underlying similarities among the people in each group, and these similarities predict what people will buy.

A system like Mosaic, with its seventy-one different lifestyle groups, can be useful to marketers. But for most people, looking at seventy-one different groups just presents a bewildering mess. In this chapter, I'm going to take an approach that I think provides a deeper understanding. Instead of supplying a prefabricated list of different types of people, I'm going to explain a few of the most important principles that give rise to lifestyle groups to begin with. Because people in a given lifestyle group tend to love the same things, understanding the differences among people that give rise to these groups provides a lot of insight into why people love the things they do. Specifically, this chapter will discuss two important differences among people — their levels of "economic capital" and "cultural capital," which lead them to belong to particular lifestyle groups and to love the same things.

Understanding economic and cultural capital brings together two of the major themes in this book. First, even though our tastes can sometimes feel inexplicably mysterious, there are predictable reasons why we love the things we do. And second, even though our tastes are intimate and personal, they are also deeply influenced by other people.

SOCIAL STATUS: A CAPITAL IDEA

Not only do I sometimes wear scuffed-up shoes, I also have this old-fashioned idea that I should keep a car until it really starts to break down. In keeping with my philosophy, I once owned a fifteen-year-old car with more than 225,000 miles on it. The damn thing just wouldn't break. Worse yet, even though the car had sustained no real damage, it still looked its age. Truth is, I felt a little embarrassed by it at times.

Once, when I was giving a lecture on status symbols, I made the mistake of using my car as an example. With what I thought was self-deprecating humor, I admitted to my students that I was sometimes unhappy when I pulled into the university parking lot and saw that all the students had nicer cars than mine. In all my years of teaching, I had never before — nor have I ever since — seen my students with such mortified faces, expressing their shocked disbelief that I would publicly admit to having such tawdry consumer-status concerns.

It reminded me of situations in which parents with young children get so used to changing diapers that they forget that for most people, graphic descriptions of their latest diaper adventures are not welcome dinner-table conversation. Like those parents, I've had my hands in people's social-status concerns for so long that I sometimes forget that, especially among highly educated people, admitting one's own social-status concerns constitutes "too much information." But the truth is that while many people do not like to admit that they are concerned about their social status, it is normal human psychology to care *a lot* about what other people think of us. And as I'll explain, that need to be admired and respected by other people influences what we fall in love with.

When sociologists discuss social status, their explanations often center around what they call capital. The word *capital* is often used to mean money. But social scientists[4] — and in particular the late French sociologist Pierre Bourdieu[5] — use the word *capital* to refer to any resource that a person can call on when needed, including their reputations and their friends. There are many different types of capital, but I'm going to focus on economic capital (money) and cultural capital (impressing people with your intelligence, sophistication, and virtue). In this view, life is, in large part, a competitive status game in which we score economic-capital points and cultural-capital points based on the things we do, say, and own.

One of the main ways the things we love say something about us is that they indicate our levels of economic and cultural capital. People score economic-capital points by owning expensive things, having high-paying jobs, hosting elaborate parties, and making big public donations to charity. That list has remained pretty stable over time, but there has been at least one significant change. In eras when wealth went hand in hand with hereditary noble titles or at least membership in landowning family dynasties, wealthy people signaled their economic capital by showing off how much leisure time they had. Hence they invented games such as cricket, whose matches last between three and five days. Today, great wealth usually comes from success in business or talent as a performer, so the more money people have, the busier they tend to be. Thus it is commonplace nowadays to hear affluent people complaining about how busy they are, which conveys how in demand they are.

Cultural capital is a lot more complex than economic capital. In the movie *Titanic*, Jack Dawson (played by Leonardo DiCaprio) is befriended by a wealthy American, Molly Brown (played by Kathy Bates). Molly Brown is considered "new money," in contrast to the "old money" aristocrats who share first-class accommodations with her on the *Titanic*. Molly says of these aristocrats, "Remember, they love money, so pretend like you own a gold mine and then you're in the club." But Molly is quite deluded on this point. Even though she has money, none of the old-money aristocrats think she is in the club. Behind her back, they call her "that vulgar Brown woman" and do their best to avoid her.

One reason Molly Brown isn't "in the club" is that she lacks cultural capital. Cultural capital includes all kinds of things that a new-money person like Molly Brown doesn't understand — the secret codes and behaviors that elude her and make her seem like she doesn't really belong in the upper class. For aristocrats in 1912, new-money people lacked cultural capital because they didn't speak with the right accent, know the fine points of formal manners, use the right words, talk about the right things, dress the right way, maintain the right posture, blush at the right things, enjoy the opera, dance the right way, read the right things, hold the right opinions—the list could go on. Cultural capital also includes owning the right objects, such as artworks that indicate your good taste, and having the right credentials, such as a degree from an elite university.

What counts as cultural capital has changed a lot since 1912. To get a quick sense of what cultural capital consists of today, take my Cultural Capital Quiz, which I created to help students understand the topic. Unlike the Love of Things Quiz (page 10), this cultural-capital scale has not been scientifically validated. However, in my humble opinion, it is still reasonably on target for the United States in 2022.

THE CULTURAL CAPITAL QUIZ

Choose whichever answer (true or false) comes closest to being correct for you. Give yourself one point each time you answer "true."

CULTURAL CAPITAL QUIZ

1.	The things I like best tend to be popular with people who are "in the know" but not popular with the mainstream public.	True	False
2.	Things that are super popular (such as the bestselling beer, number one clothing brand, and most popular TV show) tend to be mediocre.	True	False
3.	I avoid buying clothing with visible designer logos.	True	False
4.	If I were buying a new car and could afford a luxury brand such as Mercedes, Jaguar, or Cadillac, one drawback of buying a luxury brand is that I would be at least a little *embarrassed* to own it.	True	False
5.	If I had to choose between having a reputation as (a) a normal person of average intelligence or (b) a person who's brilliant but weird, I'd choose (b).	True	False

6.	I enjoy analyzing TV shows, movies, music, books, and/or art.	True	False
7.	I regularly read, watch, or listen to the *New York Times*, public radio or its podcasts, public television shows, and/or literary fiction.	True	False
8.	Instead of going on a cruise or to a big resort, I would much rather take a nature-based vacation (e.g., hiking and canoeing) or a culture-based vacation (e.g., visiting museums and historic sites).	True	False
9.	I always (or almost always) vote, even in low-profile local elections.	True	False
10.	I have changed or am changing my food habits (such as giving up meat and buying only organic vegetables) at least in part for political, ethical, or environmental reasons.	True	False
11.	I have completed (or am enrolled in) a four-year college degree program.	True	False
12.	I have completed (or am enrolled in) a four-year college degree program *and majored in the liberal arts, social sciences, or art.*	True	False
13.	I have completed (or am enrolled in) a four-year college degree program *at a private school or at the most selective public university in the state.*	True	False
14.	I have completed (or am enrolled in) a graduate degree program.	True	False
Total the number of "True" answers here:			

Scoring Guide
0–2 = Low cultural capital
3–7 = Medium cultural capital
8–14 = High cultural capital

CULTURAL CAPITAL AND WEALTH

The relationship between cultural capital and wealth is complicated. Generally speaking, cultural and economic capital go together: upper-class people tend to have lots of both kinds of capital, and working-class people often have little of either. This happens in part because children raised by affluent parents get intensive training in the skills, habits, and attitudes that will give them both economic and cultural capital as adults. And when people from less privileged backgrounds want to move up in the world by going to a university, this education increases both their earning power and their cultural capital compared to what it would have been without such an education.

But if, instead of looking at the whole population, we focus on people with a college education, we find that they face a trade-off between pursuing cultural capital and pursuing economic capital. Choosing careers in fields such as business and engineering leads to high earnings but only moderate levels of cultural capital, whereas pursuing careers in fields such as education, the arts, and journalism leads to high levels of cultural capital but only moderate levels of income.

It's not an accident that many of the fields that provide the very highest levels of cultural capital only provide low to moderate levels of income. Back in the day, it was mainly wealthy aristocrats who had the time and inclination to study topics such as literature and art, which aren't directly applicable to making money. By contrast, the middle classes worked in professions such as law, medicine, and engineering. It makes sense, then, that the aristocrats used their cultural power to establish a "rule" that expertise in non-work-related high-culture topics such as art and literature scores more cultural-capital points than work-related expertise. The vestiges of this rule still remain today.*

COUNTERFEIT GOODS YET REAL LOVE

A surefire indicator of a thing's economic- or cultural-capital status is that people try to fake it. For example, many people pretend to have economic capital by buying counterfeit luxury goods. One study[6] found that in the UK, 44 percent of consumers have intentionally bought fake designer shoes or clothing, and this does not even include widely counterfeited items such as handbags and watches. People can even rent access to movie sets that look like the interiors of private jets so they can shoot selfies that make it look like they're on a private plane.

While people fake economic capital by pretending to own expensive things, they fake cultural capital by pretending to enjoy things they don't like. For example, you might go to the symphony even though you find it dull. And if you do go to the symphony, have you ever noticed that the instant the music ends, a few people spring to their feet like Olympic sprinters at the starting gun and offer an ebullient standing ovation with almost manic enthusiasm? Perhaps this research has made me cynical, but I sometimes wonder if these audience members really loved the concert *that much* or are just trying to earn

---

* Recently, however, we've started to see a slight shift in this pattern. The dramatic rise of the tech sector and companies like Google and Apple have given technological expertise an aura of cool. Being a techie, especially if you work in a start-up that's doing something on the cutting edge, can score you a lot of cultural-capital points. But it remains to be seen if this will be a lasting change.

a little cultural capital by exaggerating the intensity of their enjoyment.

Individualism has its pros and cons, but one good thing about the increasing individualism in the United States is that there has been a marked decrease in the social pressure on people to pretend that they like high-culture stuff even if they don't. For example, expensive Cabernet Sauvignon has a reputation as one of the best red wines in the world, because a good Cabernet has the profile that wine experts enjoy: it is complex and intense and includes many subtle differences within its flavors. But that doesn't mean that nonexperts will enjoy it. Nonetheless, for many years Cabernet Sauvignon reigned as the bestselling red wine in America because wine experts said good things about it, so ordinary people believed it was the "right" kind of wine to buy. Fortunately, in the late 1990s, nonexpert wine buyers finally got up the nerve to switch from Cabernet to Merlot, which better suited their tastes and budget. Merlot quickly became the bestselling red wine in America, and as people drank the type of wine they most enjoyed, total wine sales went up.

HOW ECONOMIC AND CULTURAL CAPITAL INFLUENCE WHAT WE LOVE

The reason why the things we love *reflect* our levels of economic and cultural capital is that what we love is partly *shaped by* our levels of economic and cultural capital.

Economic capital influences what we love in two main ways. First, people *want* all sorts of things they can't afford. But wanting something and loving it are different, and people tend to love things that are part of their daily lives. True, there are some people who love things from afar — you might love Ferraris even though you've never so much as sat in one. But for every person who loves a car he doesn't own, there are lots of others who love the cars they do own. Since your level of economic capital determines what you can afford, it influences what you love.

The second way economic capital influences what we love is a bit more complicated. The higher your income, the more individualistic your identity; this, in turn, influences what you love. I explained in chapter 6 that historically, cultures gradually become more individualistic as they get richer. It is also true that at any given time within a society, the more money people have, the more individualistic they tend to be.[7]

One of the biggest differences between individualism and collectivism lies in the way people define their identities.[8] When individualistic people define their identities, they emphasize the things that make them different from other people; collectivistic people emphasize the aspects of their identities that connect them to other people. Since wealthy people tend to be individualistic, they also tend to define their identities[9] mostly in terms of the personal tastes and achievements that help differentiate them from others ("I'm an expert on such-and-such"). Middle-income and low-income people put more emphasis on the collective aspects of the identities they share with other people, such as their nationalities, neighborhoods, sports-team allegiances, and so on. This is also true, for example, when it comes to the restaurants people love. People with a high income are more likely to favor unique, newly opened restaurants that most people haven't yet tried, whereas people with a low income are likely to be regulars at a small set of favorite restaurants that they see as connecting them to their neighborhood or friends.

In terms of influencing what we love, cultural capital is even more powerful than economic capital. Creativity is one of the most important elements in today's cultural capital. As high-cultural-capital types see it, a good person is a creative person, and creative people are often a bit idiosyncratic. So being a tad odd and unconventional is a point of pride among high-cultural-capital folks, who see the word *normal* as a synonym for *mediocre*. By contrast, among moderate- and low-cultural-capital groups, being normal means being healthy, and being weird (as they would see it) is a problem. This carries over to the things each group loves. High-cultural-capital folks love things they see as creative, artsy, and unique (just like they are). Low- and medium-cultural-capital people also value creativity, but to them, a little creativity can go a long way. They don't like things that are so creative that they seem weird.

One of the main ways people acquire cultural capital is by demonstrating that they have "good taste."* But who decides what constitutes good taste? Society gives people with the biggest audiences and the most expertise in their fields the power to define good taste in their particular areas. For example, good taste in clothing is defined by "tastemakers" such as fashion writers, designers, internet influencers, retailers, and fashionista consumers. One thing all these people have in common is that

---

* I put "good taste" in quotation marks to indicate that I don't believe that anyone's taste is objectively better than anyone else's. Yet the concept of good taste is important in understanding why people with a lot of cultural capital love the things they do — and why people sometimes pretend to love things they don't.

they are fashion experts. In chapter 7, I argued that what people enjoy is closely related to their levels of expertise. The more expertise people have, the more they'll enjoy things that are challenging (for example, crossword-puzzle experts enjoy difficult crosswords). I also said in chapter 7 that foods, games, artwork, books, movies, and music are challenging to enjoy when they (1) are complex, (2) produce intense experiences, (3) require subtle discernment, and (4) require specialized knowledge to understand. Therefore, when people who want to have high levels of cultural capital strive to become experts in the things they are interested in, they start to enjoy things that are complex, intense, subtle, and a bit esoteric.

As a case in point, Cathy Horyn, writing in the *New York Times*, reviewed a runway show of fashions designed by Stefano Pilati for Yves Saint Laurent that featured the house's ready-to-wear collection — i.e., the clothing that will be sold in the mall rather than the much more unconventional haute couture outfits one often sees at fashion shows. Horyn's verdict? The clothing was "a little boring, if you really want to know."[10] This is not surprising: Horyn is a professional fashion writer with deep expertise, so her brain likes unique and innovative outfits that are visually complex, produce intense reactions, have interesting subtleties, and require a lot of background knowledge about fashion to understand. Ready-to-wear clothing is designed to please average consumers, and most things that please an average consumer bore an expert. Conversely, when typical consumers see the strange-looking avant-garde fashions that Horyn enjoys, their brains keep trying to fit these weird outfits into their understanding of clothing, as if they were trying to put a square peg into a round hole. This results in frustration and sometimes the angry conclusion that avant-garde fashions are a hoax being perpetrated on a gullible public.

This explains why the aptly named "easy listening" music is so reviled by many music lovers. Easy listening is purposely created to be extremely low in musical challenges. As a result, some people find it to be the perfect background music — something their brains can process without distracting them from whatever they're doing. By contrast, the music writer Carl Wilson[11] refers to the music preferred by music experts as "difficult listening." For these experts (and many other music lovers), easy listening is so utterly devoid of challenges that they react to it with nothing short of revulsion.

In sum, although it may seem that there are a lot of arbitrary conventions behind why loving one thing may score more cultural-capital points than loving another, there is also some method to the madness. What counts as cultural capital is often determined by experts. And because of the ways that the brain turns, say, a book into a pleasurable reading experience, experts usually enjoy things that are relatively complex, intense, and subtle and draw on background knowledge that not everybody has. When art or entertainment lacks these expert-pleasing qualities, it is likely to be seen as in bad taste, so enjoying it will score few, or even negative, cultural-capital points.

THE GREAT DEBATES

Should being an expert on classical music score more cultural-capital points than being an expert on rap? Conservative intellectuals might say yes, but lots of other people would disagree. An early skirmish in the culture wars was a debate on college campuses over whether learning the traditional canon (works by Shakespeare, Homer, and other classic writers) was a crucial part of a college education. This debate was also, implicitly, about what kind of knowledge should count as cultural capital. This is just one example of the wide-ranging debates that, while rarely using the phrase "cultural capital," are implicitly about the number of cultural-capital points a person should score for liking one thing more than another as well as the relative importance of cultural versus economic capital. And as we will see, these debates feed directly into the culture wars.

General Versus Local Cultural Capital

The debates about cultural capital often reveal a distinction between cultural capital in general and local cultural capital. When people talk about cultural capital, they normally mean things that are widely recognized across society as scoring cultural-capital points. By contrast, local cultural capital refers to things that score points within a subculture or small community. For example, among stamp collectors, knowing a lot about rare stamps scores a lot of local-cultural-capital points. In a faith community, being particularly devout can also provide local cultural capital. But in the wider society, being a stamp expert or devoutly religious doesn't earn a person much cultural capital. This often creates conflict as groups argue over what should count as cultural capital for society as a whole versus what only allows a person to be a big fish in a small pond.

Even though local cultural capital may not earn you a lot of status outside your community, it can have a huge influence on what people love. Consider, for example, the lowrider custom-car culture that emerged out of the Mexican American community in Los Angeles. (Lowriders are tricked-out customized cars that ride very low to the ground.) Creating a great lowrider car won't win you many cultural-capital points in elite circles. But our brains are automatically tuned in to notice what will win the respect of the people around us. So if you grow up in a neighborhood where the people who create lowrider cars are widely admired, your brain will notice that. As a result, you will intuitively find those cars more appealing than you otherwise might have. That doesn't necessarily mean you'll fall in love with lowrider cars, but it does give you a nudge in that direction.

When people who grow up outside high-cultural-capital circles notice that the things that score local-cultural-capital points within their communities (such as having a great lowrider car) aren't considered high cultural capital in the wider society, they can feel like their whole community is being disrespected. This leads many people to feel resentment toward high-cultural-capital elites, which adds fuel to the culture wars. Interestingly, though, people with very high levels of cultural capital have been working hard (although not always successfully) to discard the snobby attitudes that characterized previous cultural elites.

The Shift from Conservative to Liberal Values

Regarding general cultural capital, there has been a gradual shift away from conservative values and toward liberal values as a source of social status. Until the mid-1900s, being high in cultural capital meant being an aristocrat (think *Downton Abbey*). Although people now talk about "limousine liberals," the old-school aristocrats were the quintessential conservatives.

Research[12] shows that two core features of conservatism are (1) supporting traditional values, practices, and institutions and (2) being tolerant of social inequality. Aristocrats were usually conservatives because the very existence of a hereditary aristocracy depended on maintaining a traditional social order and justifying inequality. Even when aristocrats didn't have a hereditary royal title, their social status was still tied to how many generations of their family had been part of high society. In France, people even had a special vocabulary to express how long a family had been part of the bourgeoisie:* *moyenne bourgeoisie* indicated around three generations, *grande bourgeoisie* indicated around five generations, and *haute bourgeoisie* meant the family's high standing dated back to the French Revolution.

When a group has the power to decide what counts as cultural capital within a society, they use that power to bestow social status on people who reflect their values. For example, if clergy are influential within a culture, people will score status points for being pious; whereas if artists are influential, social status will go to people who are creative. Before the mid-1900s, the aristocracy had the informal power to define cultural capital. So cultural capital reflected aristocratic values: a person should be well versed in the fine arts and disdainful of popular lowbrow entertainment. It was also very important to follow an arcane system of formal manners that included arbitrary conventions such as the asparagus rule — my personal favorite — which dic-

---

* Although the French term *bourgeoisie* now refers to the middle class — historically, it denoted people of a town or city who didn't have to work at manual labor — in the United States the term also refers to many people who would be considered upper class.

tated that even at a formal dinner you were allowed to eat asparagus with your fingers.[13]

In the nineteenth and early twentieth centuries, liberal intellectuals often belonged to a tiny subculture of "bohemians"* who had their own forms of local cultural capital but were considered disreputable in mainstream society. Over the course of the twentieth century, highly educated liberals kept many of their bohemian values as they gradually came to dominate the centers of cultural power, including academia, education, the arts, and the media. With this cultural power, educated liberals have been able to redefine cultural capital in *their* own image as things that show a person to be smart, sophisticated, creative, and progressive. Today, anything that demonstrates those qualities — such as how people dress, their political opinions, their talents, where they went to college, how they decorate their homes, and what they love — increases cultural capital.

Because high cultural capital now includes having progressive political views, high-cultural-capital people have shifted from being snobs to being omnivores.[14] High-cultural-capital people used to see popular entertainment as junk that was beneath them; they were only interested in highbrow art, such as opera. Today, however, having high cultural capital means supporting progressive social values such as egalitarianism and cultural diversity. The very idea that highbrow tastes (e.g., classical music and avant-garde films) are *better* than popular tastes (e.g., Top 40 music and reality TV) smacks of elitism.

---

* *Bohémien* was the French word used to describe the Roma peoples, who were outsiders in mainstream society. This outsider status led the word to be applied to any artsy/intellectual subculture.

This places high-cultural-capital people in a dilemma: *how to show that they have superior taste without being snobby or even seeming to condone the idea that such a thing as superior taste really exists.*[15] One way they resolve this is by loving both high culture (to show their sophistication) and popular culture (to show they aren't snobs). Thus today's cultural-capital champions have become cultural omnivores who consume everything, including both public television's *Masterpiece* and Bravo's Real Housewives series. Yet when these high-cultural-capital omnivores watch, say, reality television, they still flex their cultural-capital muscles by talking about the shows in ways that show off their high levels of education.[16] This often means offering the kind of detailed cultural analysis of the shows that would be appropriate for a college seminar.

Hierarchy Is Easier to See from the Bottom

Most hierarchies are more visible to people at the bottom than they are to people at the top. For example, women notice sexism more than men do, and people of color notice racism more than whites do. This tendency for hierarchies to be most noticeable to people at the bottom plays an important role in the history of research into cultural capital. This research is most strongly associated with the French sociologist Pierre Bourdieu, who grew up in a working-class family before eventually rising to the top of French academia. As someone from a low-cultural-capital background, he could easily see how his academic colleagues, most of whom came from highly educated families, used cultural capital as a way to elevate their status. Reading between the lines in his work, I hear him saying to his fellow intellectuals, "You correctly criticize business-people for using economic capital to set up class hierarchies, but you do the same thing with cultural capital."

Similarly, the fact that cultural capital in the United States as a whole includes having liberal or progressive values goes unnoticed by some (although not all) people who share those values, but it is a frequent source of irritation to many conservatives. For example, on the *New York Times* podcast *The Argument*,[17] host Jane Coaston commented that "liberals have cultural power in America but want political power, and conservatives wield political power in America but want cultural power, so no one is happy." Her Republican guest, Michelle Cottle, elaborated, saying, "I come from a family of hard-core Republicans. . . . I get an earful about the media . . . because of [my father's] sense that [the media is] out to get him and all of his conservative friends. I think there is a real sense of persecution, or resentment, *or feeling looked down on*. And . . . he's a very affluent, well-educated Republican." In other words, her dad has plenty of economic capital and education, but being a conservative decreases his cultural capital, which leaves him feeling looked down upon by people at the top of the cultural hierarchy.

In response to feeling pushed out of the center of American culture, conservatives have been struggling to construct a counterculture with its own local cultural capital — consider Fox News, for example. Long before conservatives adopted this strategy, socially marginalized groups such as people of color and LGBTQ+ people created their own fashions, music, and local definitions of cultural capital. People who are innovators within these subcultures score local-cultural-capital points. But these status points are often discounted by the wider culture. Since people need to interact with the wider culture, this is an ongoing problem.

What's More Important, Economic or Cultural Capital?

Along with the arguments about what should count as cultural capital, there is also debate over which type of capital is more important — economic or cultural. The loudest voices in this debate come from two competing groups, the *mainstream elites* and the *cultural creatives* (see diagram on page 215). The mainstream elites consist of affluent professionals, mostly business-people but also doctors, engineers, lawyers, and others. They have an applied education that yields a lot of income and only a moderate level of cultural capital. Not surprisingly, they tend to think that economic capital should have the biggest impact on a person's social status. They argue that it is good to have money and that being professionally successful is a legitimate signal that a person is smart, hardworking, and a big contributor to society. I once heard this point made by a libertarian who argued that in capitalist economies, people are only willing to pay you a salary or buy a product from you if they think what you are offering will help them. Therefore, capitalism is a game in which you score points (i.e., earn money) by helping other people, and the person who has been most helpful to others wins. In addition to defending the importance of economic capital, mainstream elites often devalue cultural capital by arguing that it is like dessert: it's a nice optional extra, but ultimately of secondary importance. They also accuse cultural creatives of being snobs who look down on anyone who doesn't share their weird tastes.

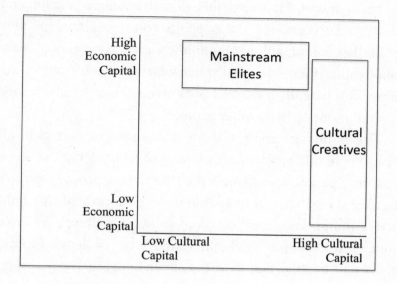

As the term implies, cultural creatives[18] tend to work in culture industries such as media, entertainment, marketing, education, journalism, arts, and social activism. Cultural creatives argue that being smart, sophisticated, creative, and socially concerned are the character traits that define a good person and hence that cultural capital, rather than income, should determine who is held in high esteem.

Cultural creatives of all income levels usually agree that awarding people high social status for having a lot of money is crass, materialistic, and unfair to those who have less of it. But the intensity of their criticism of economic capital is, perhaps not surprisingly, linked to how much money they themselves have. Cultural creatives include hipster bohemians who, like the hippies before them, have low incomes. Many of these progressive bohemians equate wealth with greed and argue that people become rich by exploiting other people and harming the environment. Therefore, they disdain economic capital and would make economic status points count *negatively* if they could. By contrast, affluent cultural creatives have enough economic capital themselves to disagree with the opinion that having money is inherently a mark of poor character, especially if they use their money to help other people.

The positions people take in this argument over cultural versus economic capital become part of their identities. Since the things people love express their identities, people with lots of cultural capital tend to genuinely love things that highlight their cultural sophistication. And because cultural creatives range across income levels, they find ways of demonstrating their sophistication that fit within their budgets. For example, high-income cultural creatives are behind much of the growing wine market in the United States, while their lower-income bohemian counterparts are more likely to become experts in microbrewed beers, which have a lot of the same complexity found in wine but are more affordable.

Similarly, people who have been financially successful are usually proud of their accomplishments and sincerely love the things that serve as trophies marking that success. I've noticed a pattern in which wealthy car collectors' most loved cars aren't the most expensive ones in their collections. Instead, they will favor the first expensive car they could afford to buy, because those cars are most connected in their minds to their feelings of success and accomplishment. People who are lucky enough to have lots of both types of capital will tend to love the things that combine them: for example, they might collect avant-garde art.

These differences between mainstream elites and cultural creatives play out when it comes to their beliefs about expensive products. Mainstream elites think that you get what you pay for, so expensive products are usually worth the cost. It makes sense for them to hold this view because it helps justify their belief that it is important to be wealthy — without wealth, you could only buy second-rate stuff. Cultural creatives, on the other hand, think that when it comes to buying the right products, the brain is mightier than the wallet. If you're smart and well informed, you can find a midpriced product that is just as good as, or better than, a high-priced brand. Therefore, high-priced brands are rip-offs that exploit the status concerns of gullible mainstream consumers. It makes sense for cultural creatives to hold this view because it undermines the value of economic capital and elevates the value of cultural capital (being smart and well informed) as the basis for being a successful consumer.

IF YOU'VE GOT IT, FLAUNT IT?

While people disagree about whether economic or cultural capital is more important, everybody cares about being respected and even admired. Economic and cultural capital can only earn you respect and admiration if other people know you have them, so we all need to display our hard-earned capital. That said, one of the biggest disagreements between lifestyle groups is over how we should go about doing this.

From the perspective of low- and middle-income people, it feels like all wealthy people flaunt their money. I spoke to an old-money guy who said that buying a $125,000 Mercedes would be showing off, so he drives a $50,000 Mercedes instead. But from the perspective of someone with less money, driving any Mercedes is showing off. As a result, low- and middle-income people see flaunting your money as the normal way rich people behave.

How low-income people respond to this, though, depends on their level of cultural capital. For people like adjunct university professors, who, despite their low incomes, have lots of cultural capital, the typical response is something like, "Flaunting money is disgusting; that's just one more reason why cultural capital is what really deserves respect." But in the more common situation, in which a low-income person also has low cultural capital, emphasizing cultural capital as an alternative to wealth doesn't help. So if these low-income people have the good fortune to someday become affluent, it seems obvious to them that they should let everyone know. As one woman who loved her $5,000 Chanel handbag told me:[19]

> It's very important to me that these bags are expensive. I've worked hard for my money, and now I want to claim the rewards. I use these bags to let people know how much money I have. And you know what my favorite part of this bag is? This part [pointing to the large logo clasp on the front].

My personal response to this changed a lot when I started interviewing people about the things they loved. I was raised to be appalled by the idea of flaunting your money. But in my research, I've also had the opportunity to speak with many people around the world who, like the woman I just quoted, were born into low-income families yet now own a Mercedes, a Rolex, a Louis Vuitton handbag, or other status symbols they cherish. When they tell the stories of their beloved items, they always begin decades before they bought them, with the poverty of their childhood homes and the years of material deprivation they endured as they clawed their way through medical school or poured every last penny into their family businesses. Eventually, they felt financially secure enough to make a big showy purchase. They tell me enthusiastically that they've always wanted whatever it is and hope everyone who sees them with it knows what they've achieved. In a research interview, it's important to keep your personal reactions to yourself, and I do. But each time I hear this type of story, instead of being privately appalled by the interviewees' crass materialism, I find myself enthusiastically rooting for them. It's easy to see why luxury items mean so much to them.

My views aside, this "If you've got it, flaunt it" attitude is completely incompatible with today's high-cultural-capital ideal of good taste. But it was not always so. Imagine if the Palace of Versailles in all its gold-leaf grandeur were being built today; high-cultural-capital folks would see it as laughably overdone.* But Versailles wasn't a nouveau riche creation; it was built by the oldest of old money. Why weren't the people who created it worried about appearing tacky?

In seventeenth-century Europe, the aristocracy had a near monopoly on both economic and cultural capital. That meant that they would win both the cultural-capital and the economic-capital competitions against other segments of society. In particular, since they were certain to win any wealth competition,

---

* This is not just a guess. Some of Donald Trump's living spaces look a lot like Versailles. And even before he went into politics, many high-cultural-capital people saw these spaces as rather grotesque.

they thought it was fine for the winners to show off their trophies (such as mansions and other seventeenth-century bling). But their attitude started to change when they were no longer assured of victory.

Increased trade in the eighteenth century followed by the Industrial Revolution in the nineteenth century meant that businesspeople were getting so rich that they could build glitzier mansions than most aristocrats could. When old-money aristocrats were losing bling contests with new-money businesspeople, how would the aristocrats respond?

The aristocracy still had the cultural power to decide how people scored points in the status game. So they changed the rules. The aristocrats might be losing economic-capital contests, but they could still count on winning the cultural-capital competition. That's because acquiring cultural capital is like learning a language: if you grew up speaking it, you'll have a level of fluency that is terribly difficult for non-native speakers to match. Old-money people are raised in high-cultural-capital families, so they speak the language of cultural capital naturally and fluently, whereas new-money people speak cultural capital as if they have had a few semesters of it in high school. So the old money started to move away from an opulent style toward a restrained style, emphasizing the idea that good taste requires simplicity and understatement. Appreciating things that are simple and understated draws on one's ability to make subtle distinctions about the way something looks, tastes, and sounds. This is particularly hard for new-money people to learn. Liking things that are elegantly understated helped the aristocracy reassert its superiority over new money.

Eventually, old-money people arrived at a compromise position in which they view it as okay to show that you have money by owning expensive yet tasteful things, but it is not okay to overtly call too much attention to possessions, since that feels like nouveau riche bragging. You can see this careful balancing act in a conversation that took place between Prince Philip and Marshal Tito in 1953. Back when there was a country called Yugoslavia, Josip Broz Tito, then the country's "president for life," had dinner with the prince at Buckingham Palace. Tito, who had grown up in a Croatian village, was impressed by all the palace finery, especially the solid gold dinner plates. He complimented Prince Philip on them. Without missing a beat, Prince Philip replied, "Oh, yes, and my wife says it really saves on breakages."[20]

DESIGNER FASHION LEAGUES

Fashion is a particularly important arena in which flaunting money plays a major role. By spending large amounts of money on prestige brands, one can score economic-capital points while sometimes losing cultural-capital points. You can see how this plays out by comparing what I call "the three leagues of designer fashion,"[21] comprising people who are enthusiastic about high-end designer brands.

The Logo League

In the Logo League, people compete by purchasing major designer products with visible logos. The bigger the logo, the more the item usually costs, hence the more points someone scores by displaying it. To stay competitive, lots of Logo Leaguers resort to buying counterfeits.[22]

The Logo Leaguers buy the *relatively* affordable products from the best-known designer brands, such as Gucci, Louis Vuitton, and Prada in the $1,000–$6,000 range; or they buy what marketers call masstige ("mass" + "prestige") brands such as Coach and Michael Kors, whose handbags typically run from $100 to $800. People in the Logo League usually have no idea that many of the rich and famous people they admire play in an entirely different league in terms of both economic and cultural capital. They are a little like Molly Brown in thinking that buying big-logo products puts them "in the club" with elite fashion lovers. I interviewed a young woman in Singapore, for example, who was far from wealthy but saved her pennies until she had enough to splurge on the bottom-of-the-line Louis Vuitton handbag (around $600 in Singapore at that time). She explained that when she used branded luxury goods with big, visible logos she felt she was "in the in crowd . . . in the thick of action, really cool. And enjoying my life while I'm still young."[23] She continued in that vein.

> **LOGO LEAGUER:** The rich and famous wear branded
> goods, especially the high-fashion type. They have
> vacations in the ski slopes, the ski resorts. They go
> to Tahiti and exotic places.
> **ME:** So when you buy branded goods . . .
> **LOGO LEAGUER:** I feel like I am a mini rich and famous
> [person] in my own right.

She clearly enjoyed the feeling she got from her purse, and I hope it continued to give her much pleasure. Yet I've spoken to people of the type she so admires, and none of them would think that her $600 purse made her part of their club.

THE LOGO LEAGUE

Gucci and Balenciaga co-branded Marmont bag, $2,800

THE LUXE LEAGUE

Hermès Verrou ostrich leather bag, $12,000

THE HAUTE COUTURE LEAGUE

Gucci Tifosa bag, $4,353

The Luxe League

Many of the truly rich and famous belong to what I call the "Luxe League." Although consumers in all three leagues buy luxury brands, the Luxe League is home to the most classic and prototypical luxury consumers. These are the people that Logo Leaguers often aspire to be. Luxe Leaguers have a lot more economic capital and, frequently, more cultural capital than the Logo Leaguers. Because they're rich and not just middle-income people making a big splurge, they can afford to buy enough pairs of $500 Gucci jeans, $800 Manolo Blahnik shoes, and $835 Prada T-shirts to wear complete designer outfits rather than just display one or two trophy pieces.

As a result, a key feature of the Luxe League is that fashion skill plays a bigger role in what people love than it does in the Logo League. The Logo Leaguers do, of course, try to stay current and assemble outfits that work. But as far as they're concerned, if you've got the right logos, you're 80 percent of the way there. By contrast, in the Luxe League, owning the right stuff is more of a given, so the focus shifts to what you do with it.

There is a competitive fashion spirit that is commonly found among serious fashionistas of all income levels. This woman's explanation for why she's not impressed by an all-black outfit in a fashion ad captures this competitive spirit:[24]

FASHIONISTA: The outfit, it's all black. Black looks good on everyone. So I really don't give anyone brownie points for being able to put together a black outfit. Yes, it looks good. But it's the simplest way to look good.

ME: So it's like they say in gymnastics: some tricks have a "higher degree of difficulty" than others. Does a person score more fashion brownie points with you if they create an outfit with a high degree of difficulty?

FASHIONISTA: Mm-hmm. I would say that's very accurate. If you put on a cute graphic shirt and jeans, anybody can do that. The outfit did the work for you. Whereas if I'm wearing a cool print on my jeans, that's, like, really hard to match. And I put on some shoes and my hair and everything works together. You have to have some skill to put that together and have it look nice. Then I'm really impressed with somebody that does that. Versus if you just put on all black — like, that's easy.

The Luxe League combines that fashionista attitude with a *Lifestyles of the Rich and Famous*–level income. Luxe Leaguers love the big established luxury brands such as Chanel, Gucci, and the rest but sometimes feel conflicted about logos. On the one hand, they want to show good taste and distinguish themselves from Logo Leaguers by avoiding giant designer logos that are visible from across the room. But they still want the people they're talking with to appreciate what they have. So they often choose items with small logos or items such as the famous Birkin bag, which comes in styles ranging from $40,000 to $500,000 for the most desirable collectibles[25] and is so iconic that it will surely be recognized by people who are in the know.

The Haute Couture League

The Haute Couture League is a small group of global superelites who sit at the top of the fashion status hierarchy. When it comes to wealth, they are often the 0.1 percenters. But more important, they are usually trust-fund kids who grew up rich. Earlier in this chapter, I compared learning the skills that produce cultural capital to learning a language. Typical Haute Couture Leaguers are not just high-cultural-capital native speakers; they also grew up in homes where luxury brands were lying around the house — the normal stuff. As a result, they feel very at home with these brands. And they don't hold the same type of reverence for mainstream luxury brands that consumers in the other leagues do.

Haute Couture Leaguers live in cities such as New York, London, Hong Kong, Tokyo, Milan, and Paris, and they fancy themselves not only wealthy but also members of the cultural avant-garde. Because Haute Couture Leaguers are high in cultural capital, they value creativity and a willingness to defy social convention. It's easy to identify a fashion ad targeting this group. Just show the ad to a typical consumer. If the consumer responds, "That outfit's so weird — would anyone actually wear that?," the ad is targeting the Haute Couture League.

Members of the Haute Couture League buy plenty of expensive stuff, but they don't want to be confused with Luxe Leaguers or even, God forbid, Logo Leaguers. So although they sometimes buy from the big designer brands, they often buy the brand's quirkiest products — such as the Gucci Tifosa bag pictured on page 223 — and avoid the mainstream favorites. Similarly, they buy from less well-known brands, such as Bottega Veneta or even more obscure alternatives such as Pyer Moss or Stephane Rolland. This sends the message that they are so above it all that they don't care if the rest of us know what brands they buy.

Haute Couture Leaguers show their mastery of fashion by having a playful, even flippant, attitude about it all. One woman I interviewed had an Haute Couture sensibility yet was surrounded by Luxe and Logo Leaguers, and she didn't much like it.[26] For sport, she would troll the Luxe and Logo Leaguers by wearing items from famous brands that were tasteful but recognizable as very expensive (around $20,000 or more for an outfit). She would also carry a supercheap, obviously fake designer purse. The outfit as a whole said, "I clearly have enough money to buy a real designer purse, but I'm using this cheap fake anyway." She enjoyed the confusion and discomfort the outfit created.

THIS CHAPTER EXAMINES what the things we love say about us. These things frequently identify us as belonging to a particular lifestyle group. These lifestyle groups differ from one another in how much they value economic versus cultural capital and in how much of each type of capital their members typically have. These groups also differ in how they define local cultural capital, meaning the beliefs, actions, and possessions that prove a person is living up to the ideals of that group's particular subculture.

Economic and cultural capital also influence what we love. To borrow a phrase from Marie Kondo, we know we love something when it "sparks joy." These sparks of joy are generated by our unconscious minds. I've argued throughout this book that they occur for several reasons, one of which is that our brains think these objects will score economic- or cultural-capital points with the people we care about. In that way, the influence of economic and cultural capital on what we love can be sizable, yet we might not recognize it for what it is, because it simply feels like a mysterious attraction to a particular object or activity.

9

Because Evolution

WHEN WE SAY SOMEONE IS "MONEY HUNGRY," WE MAY BE MORE accurate than we know. David Gal,[1] a professor of marketing, reasoned that since money represents power and status, it would be particularly appealing to people who felt powerless, so he conducted a study to test his idea. To whet the appetites of the subjects of his study, Gal asked them to write about a time when they lacked power. Once they were in this somewhat needy frame of mind, he showed half of them a picture of money and the other half a picture of office supplies. He and his colleagues then measured the salivary responses from all the participants. What happened? The people who were shown the picture of money salivated, whereas the office-supply gazers did not. Salivating before eating food makes sense because it helps with digestion, but why would people salivate over a picture of money?

Like most things in the modern world, money didn't exist at the time our bodies evolved, so we don't have an evolutionary response specifically for money. In effect, our brains are saying, "I'm not set up to deal with money, but money is kind of like food in that both are important resources. So I'll borrow some behaviors that evolved for food and apply them to money." Your brain is able to modify these behaviors a bit; after all, we don't eat the money. But many food-related behaviors, such as salivation, are triggered by other things we desire. In chapter 3, I labeled these phenomena "carryover effects," because responses that evolved for one situation — in this case, eating — carry over to a different situation, where they create some odd behaviors.

Throughout this book, I've cited lots of specific examples of carryover effects. But if we take a big-picture view, the whole phenomenon of loving things can be seen as one giant carryover effect. Human love evolved over millions of years to create relationships between people. But loving things didn't become a major part of our lives until long after our brains had evolved to their current state. When we love things, our brains borrow a prehistoric psychological mechanism that evolved for human relationships (love) and carry it over, with some adjustments, to our relationships with things.

In the previous chapters, I've focused mostly on how the love of things works. But if we want to understand not just how our love of things works but also *why* it works that way, we have to look all the way back to the evolution of love. Doing so means exploring new topics while providing a much deeper look at key points made earlier in this book. So this chapter will provide a review of some of the book's main themes.

THE EXPANDING CIRCLES OF LOVE

To understand the love of things, it's helpful to break its development into four stages. The first three stages are based on physical changes in the brain.[2] The fourth stage reflects the cultural change discussed in chapter 6, as Romantic ideas about love gained popularity and things became plentiful.

STAGE 1:
SELF-LOVE

There are moments in raising a child — say, during a toddler's classic screaming meltdown in the supermarket — when a parent could be forgiven for envying a fish. Many female fish lay lots of eggs at one time. Then the male fertilizes them, and, having completed all their parental duties, both the male and female swim merrily away.* This is what scientists call "an *r* reproductive strategy": have scads of offspring so that at least a few of them will make it to adulthood, but don't invest your time or energy in raising them. What's more, for an *r* strategy to be successful, animals don't need a psychological mechanism that compels them to take care of their offspring; they just need a sex drive that points them toward a physically healthy partner. In other words, as the Beatles didn't say, all they need is lust.

This type of lust is related to, but quite different from, love. Sexual attraction helps fuel romantic love, but it is not normally a part of parent-child love or love between friends. Perhaps for this reason, research[3] into the ways most people define love

---

* Or, in some species, die, but let's not dwell on that.

shows that nonsexual aspects of love, such as caring for and respecting the love object, are far more central to the way most people understand love than is sexuality. The fact that sexuality isn't a necessary part of love per se explains why we can love all sorts of things without feeling sexually attracted to them.

That said, there are still some underlying psychological similarities between our lust for potential mates and the attraction we feel for things we love. To begin with, lust and our love of things are both easily triggered by good looks. It's not uncommon to feel sexual attraction toward a really hot person, even if he or she is a bit of a jerk. This makes evolutionary sense, because this kind of sexual attraction first developed in animal species that didn't feed and protect their children. What difference does it make if that sexy potential mate is not a paragon of responsibility, since he or she is going to swim off the moment mating is over anyway? The only thing your mate will give the children is genes, so that's all you need to worry about. It makes sense, then, that the physical traits we are attracted to in a person, such as athletic fitness, unblemished skin, and even facial symmetry (the left half of your face looks like the right half), all indicate good health and good genes.[4]

Just as beautiful people trigger sexual attraction, so, too, can beautiful things — cars, shoes, and other objects — trigger a lusty response. Consider the passionate attraction one man I interviewed felt for some clothing he liked the look of.[5]

> I saw [an outfit], and it's like one of those things when you're in a store, and you see something and it screams your name, you have to buy it. That's the way I feel about these clothes.

Another thing that good-looking people and good-looking things have in common is that they can both have an unusually powerful impact on the way we feel about ourselves. If we acquire a good-looking romantic partner, that tends to boost our self-esteem. Marketing professors Claudia Townsend and Sanjay Sood[6] have also shown that if we acquire a good-looking product, this boosts our self-image more than getting a functional but unattractive product does. Therefore, when people are feeling insecure, they feel the need for self-esteem, which in turn increases their attraction to good-looking products.

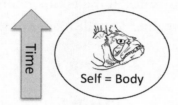

Stage 1: The self and love are limited to the body.

Animals that use an *r* reproductive strategy, in which mates and offspring fend for themselves, haven't evolved to love their progeny. But these animals, and indeed all animals, have what the eighteenth-century philosopher Jean-Jacques Rousseau[7] called self-love. As he wrote, "Self-love is a natural sentiment which prompts every animal to watch over its own conservation." This self-love is a collection of brain mechanisms that prompt animals to feed and protect themselves. Logically, these brain mechanisms must include a basic self-concept that tells them "This is you," so they won't try to eat their own tails for lunch. Furthermore, these brain mechanisms treat everything within the self (i.e., an animal's body) as intrinsically important, meaning it is always important; everything else is treated as extrinsically important, meaning it is only important to the extent that it potentially threatens or benefits the self.

At this point you may have noticed that some of the key findings I discussed earlier in the book are starting to appear in the evolutionary story. Animals have a self-concept that draws a boundary between their bodies and everything else. This self-concept also becomes a boundary between what is loved (themselves) and what is not loved (everything else). Finally, their brains treat things that are loved (themselves) as intrinsically important. When animals start bonding with their offspring and mates, these components of love will start fitting together, creating something that resembles love as we know it.

STAGE 2:
THE CIRCLE OF LOVE EXPANDS TO THE FAMILY

Most pet owners believe that their pets love them. But until recently, the mainstream scientific view was that although some animals may do things that seem to indicate their love for us, they aren't capable of love as we know it. However, there has been a general scientific trend in the other direction: the more we learn about animals, including their emotional bonds, the more like us we discover them to be. We now have brain-scanning technology that allows us to look beneath the surface of animal behavior. Much of this research has been conducted with prairie voles (animals similar to gerbils), which mate for life. Using brain-scanning technology, we have learned that when a prairie vole comes into contact with its mate, its brain activity looks the same as that of a human brain experiencing love.[8] That's strong evidence that these mammals are experiencing something that can legitimately be called love. It's probably safe to assume that vole love is different from human love, just as a vole's hunger or sleepiness is probably different from the human versions of these things. But we don't insist that a vole's hunger needs to be identical to a human's in order to qualify as hunger, and the same should go for love.

One of the reasons vole brain scans can look like human brain scans is that both voles and people are mammals and therefore have structurally similar brains. A lot of the neural activity known as love takes place in a part of the brain called the neocortex. Only mammals have a neocortex, so only mammals are capable of full-on love. Yet some nonmammals, such as many bird species, display behaviors that look a lot like love. When a male bird is trying to woo a mate, for example, his "attraction is regularly associated with heightened energy, focused attention, obsessive following, sleeplessness, loss of appetite, possessive 'mate guarding,' affiliative gestures, goal-oriented courtship behaviors, and intense motivation to win a specific mating partner," according to one neurological study.[9] Sound familiar? Scientists call this lovelike attraction between mates pair-bonding. In addition to this pair-bonding between mates, many animal species also form bonds with their young.

It's not difficult to understand why bonding within an animal family could give it an evolutionary advantage. The alternative to an *r* strategy is a *K* strategy,* in which animals have few

---

* Yes, *r* is lowercase and *K* is uppercase. I'm told it makes sense if you speak German.

offspring but invest time and energy in parenting them. Neither an *r* nor a *K* strategy is inherently better than the other. What's best depends on the type of animal (for example, *K* strategies work better with large animals) and its environment (*r* strategies work particularly well underwater). One theory is that when the first animals emerged from the oceans and started laying eggs on land, a *K* strategy, in which parents raise their offspring, became more advantageous. And if one animal parent is going to tend to the nest, it's often helpful if the other parent brings home food for everyone. This is why bonding between mates made sense for some species. But for animals to engage in any of this "caring for others" behavior, they needed to evolve a motivational system in their brains that would produce the behaviors. In animals, we call this motivational system bonding. In some mammals, bonding would later evolve into love.

This may explain why bonding evolved in some animals, but it doesn't explain how this new motivational system within the brain came into being. What happened to make bonding, and then love, possible? Because evolutionary changes first emerge as random mutations, it is much more likely that something that already existed underwent a small change than that something completely new popped into existence out of nothing. David Linden, a professor of neuroscience, put it well: "[In] evolution, you never build something new if you can adapt something you've already got."[10] So what did these animals already have that could have been adapted to produce these parenting behaviors?

All these animals had self-love, which caused them to care for themselves. All they needed was for their self-concepts to

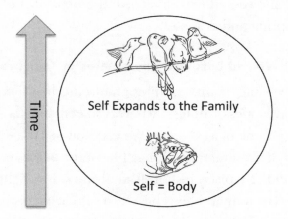

Stage 2: The self and love expand to include the family.

expand, as depicted in the diagram above.* That one simple change would lead parents to take care of their children and/or mates, thus opening the door to all the behaviors that are part of animal bonding and human love. In the diagram, the circle can be thought of as a "circle of self" as well as a "circle of love." As the circle expands from containing only the animal's body to including the animal's family, this indicates that the animal has evolved the ability to expand its sense of self beyond its previous boundaries and, in so doing, bond with (and later love) things other than itself.

In sum, what I am proposing here is that the reason human love involves expanding the self to include the love object is that long ago, this process of self-expansion was an easy evolutionary

---

* Psychologists and biologists see things from different disciplinary perspectives. As a psychologist, I talk about changing mental states, whereas evolutionary biologists would describe these events in terms of brain structures, neural connections, and hormonal activity.

step that allowed animals to evolve the behaviors of caring for their offspring and/or mates.

Love and Lust Are Roommates in Your Brain

I said in chapter 3 that exciting brands are like the people we think of as hotties or flings,[11] whereas sincere brands are like the people we think of as keepers. We can now see the evolutionary origin of that distinction. When long-term pair-bonding developed, it didn't replace lust; instead, it moved in right alongside it.[12] So today, human brains have two somewhat distinct systems that create attraction: (1) a sexual attraction system that looks for hotties and (2) a long-term love system that attracts us to keepers. These two systems are not entirely separate: having sex leads to the release of hormones in the brain that promote long-term bonding with one's partner.[13] But these two systems can also act fairly independently of each other. For example, Helen Fisher, an anthropologist at Rutgers University, and her colleagues[14] have pointed out that you can activate the sexual desire system in both men and women by giving them testosterone, but doing this only increases lust; it doesn't increase love for a partner. If you've ever been interested in two potential partners, one who is a hottie and the other who is a keeper, it may have felt like you had two distinct voices in your head, each telling you to do something different. It felt that way because that's pretty much what was going on.

The distinction between hotties and keepers also has an analogue in the kinds of foods people love — or, more specifically, do not love. When I started asking people to tell me what they loved, I assumed that lots of people would mention desserts. But it turns out that while many people are attracted to desserts, they rarely love them. In one of my studies, twenty-two people named a food they loved, but in only three instances was this food a dessert.[15] Furthermore, two of those three people changed their minds later in the interview and said they didn't truly love the desserts. Why do desserts get so little love?

The two people who changed their minds said the problem was that the desserts weren't healthful. The desserts were just flings — attractive and enjoyable but not suited for real love. What of the one last person who did love a dessert? The dessert she loved was frozen yogurt because it had the "taste without the guilt." To her, frozen yogurt was that rarest of finds: a hottie that is also a keeper. This also brings to mind a woman I spoke to in Switzerland who told me how much she truly loves chocolate and that, in the Swiss view, chocolate is good for your health (she saw it as a hottie and a keeper, too). This tendency to be attracted to, but not love, things that we see as harmful also is evident in our relationships with cigarettes, alcohol, drugs, and lots of other stuff. When we see things as attractive but harmful, we may "lust" after them, but we're much more likely to love them if they are both attractive and good for us. That said, if we are in the opposite situation, in which something is seen as healthful but bland, that doesn't inspire love, either.

When it comes to people, we'd also like our romantic relationships to be with partners who are both hotties and keepers. Yet we don't always get that. As the noted sociologist Bernard Murstein wrote, "Only individuals with numerous interpersonal assets and few liabilities really *choose* each other. Those with fewer assets and more liabilities often *settle* for each other."[16] Ouch. To be clear, I'm sure that people who have the intelligence and impeccable taste to be reading this book are so attractive that they don't need to "settle" for their partners.

But just in case you might know someone who has been in this position, I'd like to offer this supportive thought. Settling is a good thing. Settling shows that the need for love is so powerful and profound that it can overpower our concerns about how good-looking someone is, how much money he or she has, and other superficial characteristics. And that's one of humanity's most redeeming traits.

An Alternative Theory About What Came First

If love were a TV program, *Loving People* would be the original hit show, and *Loving Things* would be the spin-off. It's much more common for a behavior to have developed for interpersonal relationships and then spill over onto our relationships with things than it is for a behavior that developed for things to spill over onto our relationships with people (although I have written about how this can happen in our dating relationships).[17] I sometimes call this phenomenon "interpersonal primacy," meaning that our relationships with people came first and created the template for what love is, then the template carried over to our relationships with things.

However, noted biologist Larry Young has a different theory: some animals first evolved an attachment to things that then became the template for their attachment to their mates. Young studies the evolution of pair-bonding among voles.* In some species of voles, the males are territorial, meaning that they see a certain territory as "theirs" and will defend it against intruders

---

* Yes, voles again! Voles are the animal superstars of research into the biochemistry of love and attachment. They are often used in research because one type of vole, the prairie vole, forms strong pair-bonds with its mate, whereas other types of voles, such as montane voles, do not.

who would compete with them for food. Young's theory is that these male* voles first evolved the tendency to bond with a territory, which created what in people we'd call "psychological ownership" of the territory. Later in their evolutionary history, their brains evolved to extend this concept of owning a territory to, in effect, owning their mates. Seeing one's mate as property is not a very romantic thought. But there is evidence in the work of Young and his colleagues,[18] mostly having to do with the way hormones function in a vole's brain, that supports this theory. His theory also explains an odd bit of male vole behavior. When a male vole pair-bonds with a female, he will prevent both male *and female* voles from getting too close to her. Keeping rival male voles away from his mate makes sense. But keeping female voles away . . . not so much. Young suggests that this odd male behavior could be a holdover from the time when the behavior first evolved — say, as a way to guard territory from both male and female voles that might eat the food there. However, interesting as this possibility is, we are still a long way from knowing if it is true for voles, much less whether it is also true for people. So when it comes to people, I think our best guess remains that love evolved first for other humans and was only later applied to things.

STAGE 3:
THE CIRCLE OF LOVE EXPANDS TO THE GROUP

A common theme throughout this book has been the many ways our relationships with objects are connected to our relationships

---

* There is evidence that if this theory is correct, it applies only to male voles pair-bonding with females, not female voles pair-bonding with males.

with people. Our love objects help us have fun with other people, give us something to talk about with them, remind us of our friendships, strengthen our group identities, help us gain other people's respect, and so on. To understand why people play such a big role in our relationships with things, we need to understand the evolution of friendship, which starts with the neocortex.

The neocortex is essential for love and many other aspects of social relationships. In humans, the neocortex evolved to be very large, making up around three-quarters of the brain. The fact that humans have the largest brains (relative to body size) of any animal is mostly because we have such hefty neocortexes.

Having a large neocortex is in some ways a *dis*advantage. Large brains require large heads, which cause some infants and mothers to die in childbirth. What's more, our brains use eight to ten times as many calories per unit of mass than does skeletal muscle.[19] So even though our brains make up only 2 percent of our body weight, they use up to 20 percent of our daily calories.[20] That means our large brains require us to find a lot of food. For our ancestors to have evolved such a big neocortex, it must have provided advantages that more than compensated for these disadvantages.

There is a popular misconception that the main advantage of our big brains is that they allowed us to make tools. Scientists now doubt this explanation, because during the time when humans were evolving large brains, our tools didn't change much. To understand the main reason why we have big brains, it helps to start by comparing the brain sizes of various animal species. Species that mate monogamously and parent their offspring have significantly larger brains than species that don't do these things.[21] Biologists believe these large-brained species need that extra brainpower in order to make smart choices in mates and to coordinate parental teamwork as they take care of their children. So in all sorts of animal species, the main purpose of a large brain is to allow for complex and cooperative relationships between parents.[22]

Humans* took this whole thing a step further. During the period when our neocortexes were growing rapidly, we didn't just coordinate parenting with our mates; we also coordinated many aspects of daily life with other members of our group who were not part of our immediate families. Robin Dunbar is perhaps the best-known researcher of the ways in which primates' social needs have affected the evolution of their brains. He has shown that if you compare various species of apes and monkeys, species that have big brains also live in large groups and have complex and flexible social organizations within these groups. This suggests that in humans, what started out as love-based cooperation within a nuclear family later extended to the tribe. This was good for our evolutionary success because groups of animals are teams that compete with other animals for resources. And big, flexibly organized teams tend to trounce small, rigidly organized teams.

Every good coach knows that to create an effective team, you need to motivate teammates to care about one another. Or, to paraphrase the famous biologist E. O. Wilson, a selfish individual will beat an altruistic individual, but a team of altruists will beat a team of egoists. Therefore, for humans to function effectively in teams that extend beyond the immediate family,

---

* Apes and monkeys also show some of these same evolutionary patterns. And some other animals seem to have friendships outside their nuclear families.

we evolved the ability to care about people beyond our nuclear families. In other words, we evolved the capacity for friendship. Although we don't always talk about friendship this way, close friendships are a form of love. Love is a big part of what makes intimate relationships of all kinds feel close. Research conducted by Arthur Aron and others has shown that not just romantic love but also close relationships in general work by including other people in one's sense of self.[23] This is hardly a new idea. As Aristotle wrote: "A friend is a second self."

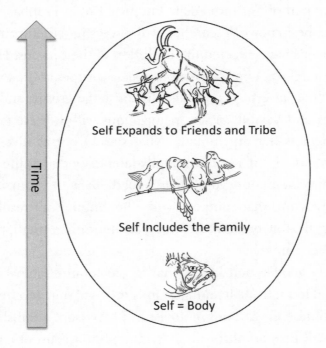

Stage 3: The self and love expand to friends and to the tribe.

I believe our ancestors acquired the capacity for friendship the same way they acquired the capacity for emotional bonds with their children and mates: their brains evolved in such a way that, under the right circumstances, their sense of self would enlarge to include individuals beyond their families. As this happened, the circle of love expanded yet again.

Since a sense of identity can include parents, children, other relatives, and friends, an individual's self-concept might hold enough people to start a small orchestra. We see evidence of this in research conducted by Stephen Dollinger and Stephanie Clancy of Southern Illinois University,[24] who asked college students to take twelve photographs that would capture "how you see yourself." The most common *self-descriptive* photographs were photos of *other people* — 98 percent of the respondents brought in at least one photo of another person, whereas only 84 percent brought in one or more photos of themselves.

Not only did large brains give us the social intelligence to work together as groups, but those of us with big brains were also more likely to succeed and reproduce within those groups. Many people misunderstand human evolutionary history. When they think about people's lives in prehistoric times, they assume that our biggest challenges were things like avoiding saber-toothed tigers — as opposed to people's lives in the modern era, when our biggest challenges involve interpersonal issues such as dealing with difficult bosses. I will admit that the number of people killed by saber-toothed tigers has declined markedly since that species became extinct. But even back in prehistoric times, one's fate depended on one's social skills. First, just as they do today, one's social skills had a lot to do with one's ability to find a mate. Second, today's machines, from tractors to kitchen appliances, help us get things done. But before the advent of technology, the most powerful and versatile resources for getting things done were friends. Third, social relationships with other tribe members had a big impact on people's roles within the group and their allotted share of the collective resources. And fourth, today, if you stink up your social relationships at work, you can always find a new job. But in prehistoric times, you saw the same people all day every day. It was difficult to escape a relationship gone sour. All this led our ancestors to evolve brains that would help them succeed in relationships with other people.

Because the human brain, and particularly the neocortex, evolved largely to help us succeed at social relationships, we have what many scientists call a "social brain." I mentioned the social brain thesis briefly in chapter 1, but it requires a little more attention. Scientists used to view the brain as a general-purpose device that could think equally well about people, rocks, or whatever. But now, biologists have identified many areas of the brain that have specifically evolved for thinking about people. For example, one part of the brain, called the "fusiform face area,"[25] evolved specifically to help people recognize human faces. A striking example of the importance of this was described in an article in the *Journal of Neuroscience*.[26] Ron Blackwell, a patient who underwent brain surgery to reduce severe epileptic seizures, received mild electrical stimulation to various regions of his brain so doctors could locate the part responsible for his seizures. The patient remained awake during this process (I know — yikes) so he could report the effects of the electrical stimulation to the surgeon. When the surgeon stimulated the fusiform face area, Blackwell reported that the surgeon's "whole face just sort of metamorphosed," giving him the impression that the surgeon had "just turned into somebody else." Yet no other parts of the surgeon's body, or any of the other objects in the room, were affected.

The social brain thesis also explains why we enjoy being generous. A study conducted by psychologist Lara Aknin showed that even toddlers who were too young to have been taught to share smiled more when sharing a treat than when receiving a gift. In another less scientific but much cuter example, my wife, who is a rabbi, was putting on a puppet show for the preschoolers in her congregation. One puppet held a stuffed animal. The other puppet asked if he would share, but the first puppet refused. When the puppet without a stuffed animal started sniffling and looking sad, a two-year-old boy got up from the audience, toddled over to the puppets, and handed his stuffed cat to the sad puppet. Just as we evolved to enjoy behaviors such as eating because they are important for survival, we enjoy being generous because it helps foster social relationships, and in the group-focused environment in which we evolved, social relationships, too, were a survival necessity.

IT HAS LONG BEEN SAID that humans are social animals. But only recently have biologists and neuroscientists begun to demonstrate the extent to which our social nature is hardwired into our brains. *Even when we relate to things, we do so with a brain that prioritizes people.* This is why, as I noted in chapter 4, Belk's first axiom states that relationships that at first appear to be person-thing on closer inspection so often turn out to be person-thing-person.

Why Our Brains Sort People from Things

This book has been organized around what I call the three relationship warmers: anthropomorphism, people connectors, and inclusion within the sense of self. These relationship warmers are ways that cool, pragmatic relationships with things can be "warmed up" and qualify as love. All three relationship warmers get your brain to think about things in ways that it normally reserves for people.

But why does your brain usually think about things and people in different ways, sometimes even using separate neural systems for people as opposed to things? As researchers Nicolas Kervyn, Susan Fiske, and Chris Malone[27] report, "The medial-prefrontal cortex is activated when someone looks at a person performing a task, but it is not activated when watching a robot perform the exact same task." Another team of researchers — Carolyn Yoon, Angela Gutchess, Fred Feinberg, and Thad Polk[28] — found that we process information about people's personalities primarily in the medial prefrontal cortex, but when we think about "brand personality" (e.g., whether the Apple brand is exciting or sophisticated), information is processed mainly in a different brain region, the left inferior prefrontal cortex, where we tend to think about objects.

Why does the brain often think about people and things in different regions? Earlier in this chapter, I described the social brain thesis, which holds that our brains evolved to meet the challenges of living with other people, and that sometimes led our brains to evolve specialized mechanisms for thinking about people. One theory takes this a little further and says that the mechanisms for thinking about people are, in general, more powerful than the mechanisms for thinking about things. Evidence supporting this view comes from a logic problem called the "Wason selection task." This task involves four cards laid out on a table, each with a number on one side and a pattern, either dots or bricks, on the other. Which card(s) must you turn over to test the proposition that *if a card shows an even number on one face, then its opposite face shows dots*? Your goal is to answer the question after turning over as few cards as possible.

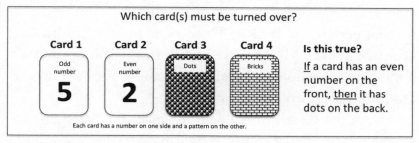

Wason selection task 1

If you're like most people, you said you would turn over the card showing the number 2, the card with dots on the back, and maybe the card with bricks on the back. If so, like more than 90 percent of people who have tried this puzzle, you'd be wrong. I'll explain why in a moment, but first here's another logic puzzle to try.

You are the chaperone at a party. People younger than the age of twenty-one can only have soft drinks. You see four people. Which people do you need to investigate further to see if the rule is being followed?

You can see that person 1 is thirty years old and that person 2 is sixteen years old. They both have drinks, but you can't tell what they are drinking. Person 3 and person 4 have their backs to you, so you can't see how old they are. But you can

Wason selection task 2

see that person 3 is drinking Coke while person 4 is drinking beer (see above). Most people find this easy. You need to see if the sixteen-year-old is drinking beer and if person 4, who is drinking beer, is over twenty-one.

Here's the punch line. Logically, these two questions are identical. The correct answer is the same for both: you want to investigate cards 2 and 4 and persons 2 and 4. Why is the party question so much easier to answer correctly? This logic puzzle has been called "the single most investigated experimental paradigm in the psychology of reasoning,"[29] so people have looked into it a lot. Researchers have found that people have a much easier time answering the question when it is about people than when it is about objects.[30] When the problem is about people, we instantly translate the statement "People under the age of twenty-one can only have soft drinks" into the logically identical question "Is anyone under twenty-one drinking beer?" From there it's easy to see that we need to investigate the person under twenty-one and the person drinking beer. But our brains have a much harder time translating the rule "If a card shows an even number on one face, then its opposite face has dots" into the question "Are there any cards that have an even number on the front and bricks on the back?" Once you've made that translation, it becomes clear that you only need to look at cards that have an even number on the front (card 2) and bricks on the back (card 4). The fact that this logic puzzle is so much easier to solve when it is about people than about things is seen as evidence that the parts of our brain that are focused on people perform at a higher level than the parts that think about things.

What does all this have to do with relationship warmers? One of your brain's most frequently performed tasks is to screen out information that it doesn't have the bandwidth to deal with. Social psychologist Timothy D. Wilson[31] notes that every second, we receive around eleven million bits of information through our senses, yet our brains are only able to process around fifty bits of that information. So your brain uses a variety of sorting mechanisms to reduce the incoming information to a manageable level. One thing those sorting mechanisms do is separate information about people from information about things and make sure the mental capacities (such as love) that evolved for thinking about people don't get bogged down thinking about all the millions of things that pass before our eyes every day. The three relationship warmers fool the sorting mechanisms into treating certain things as if they were people. Anthropomorphism, the first relationship warmer, "disguises" things as people. The second relationship warmer, people connectors, associates things with certain people so that the sorting mechanism sees them as human. The third relationship warmer is a lot like the second, but instead of associating things with other people, it associates them with you. They become part of your identity, your sense of self.

Relationship warmers make things eligible for love, but they don't guarantee love. There are plenty of anthropomorphic things that we don't love. The same is true of things that we strongly associate with other people. For example, you may associate your aunt's tomato soup with her, which leads you to feel more emotionally connected to it than you do to the can of tomato soup on your shelf. But if you think your aunt's secret recipe is so bland that it ought to remain secret forever, you're not going to love her soup. And even if we see something as part of ourselves, we all have bad habits and other things about ourselves that we don't like, much less love. Relationship warmers can make loving a thing possible, but the thing still needs to please you, feel right to you, and leap over a number of other high hurdles to become loved.

STAGE 4:
THE CIRCLE OF LOVE EXPANDS
TO PRETTY MUCH ANYTHING

I said earlier in this chapter that prehistoric animals had what has been called a kind of self-love that motivated them to care for themselves. Later, some species expanded this to include the family; and later still, some expanded it again to include friends. But when did the ability to love things develop?

Do Animals Love Things?

If some animals love things, that would support the idea that the ability to love things may have developed early in our evolutionary history, perhaps even before we became human. The phenomenon of brood parasites — animals that trick other ani-

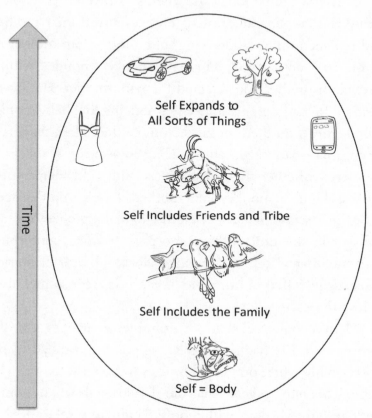

Stage 4: The self and love expand to all sorts of things.

mals into raising their young — is relevant here. For example, the European common cuckoo finds the nests of other bird species and lays its eggs in them. The mother bird whose nest has been invaded usually accepts the extra egg as one of her own and later bonds with and feeds the baby cuckoo. The invasive cuckoo isn't a thing, but it also isn't one of the mother bird's offspring. So there is at least some flexibility in what animals can bond with. Does this flexibility extend to bonding with things?

Consider the famous experiments[32] in which psychologist Harry Harlow separated infant monkeys from their mothers* and replaced the mothers with both a terry-cloth mother doll and a wire mother doll. The baby monkeys bonded with the terry-cloth mother and rejected the wire mother. Harlow was inspired to conduct this study after noticing that baby monkeys raised away from their mothers clung to their cloth diapers the way human babies sometimes cling to security blankets. He wondered whether the soft texture of the cloth diaper might remind the baby monkeys of a mother's touch, and that seems to be the case. Adding a terry-cloth skin to the monkey's mother doll made the doll anthropomorphic (monkeypomorphic?) and encouraged the baby monkey to bond with it. Harlow's work suggests that in principle, it is possible for some animals to love things.

Many people feel that, for example, their dogs love their favorite toys. My two dogs get along great, but they can fight occasionally when one of them tries to take the other's chew bone. This *may* indicate that the dog loves this bone, but we can't be sure. My dogs' willingness to fight for something shows that they value it. But loving something includes having an emotional concern for it that goes beyond valuing it for the benefits it provides. I'm not aware of any research that would tell us if animals love objects in this way or if they simply value them because they provide nutrition, comfort, or pleasure. For now, the most we can say is that some animals seem to have the capacity to love things, but we don't know for sure if they do.

---

* While these experiments were scientifically important, it's hard to read about them without feeling terrible for the monkeys.

When Did Humans Start Loving Things?

Archaeologists have found engraved shells created by *Homo erectus* around five hundred thousand years ago. Because people often love their own creations, it's plausible that our early human ancestors might have loved such things, although their brains still had some evolving to do and we don't know if they were capable of forming the same kinds of emotional attachments to objects that we do today.

Humans evolved fully modern brains around fifty thousand years ago, so the biological capacity to love things goes back at least that far. What might these early humans have loved? Archaeologists have found significant artwork going back forty thousand years, and a few finds go back even further than that. That artwork includes idols, which are excellent candidates for love since they are highly anthropomorphized objects. And when people today pray to religious figurines, they often talk to the figurines, petitioning them for some favor, which encourages a sense of relationship with the object. Ancient humans may also have loved nature, which, as I noted in chapter 1, is probably the most widely loved thing today.

Aside from idols, did early humans love their other possessions? Aboriginal tribes give us useful clues about what early human life was like. Bronislaw Malinowski,[33] the founder of modern anthropology, wrote about the Kula gift exchange system used by some of the indigenous peoples in New Guinea. Kula gifts are exchanged between high-status people in various communities, and gift givers often undertake dangerous journeys of hundreds of miles by canoe in order to reach the people they are giving gifts to. Gifts are always either necklaces or armbands made from shells, and if one person gives a neck-lace, the recipient must reciprocate with the gift of an arm-band. Some Kula objects are more prestigious than others: the prestige value increases as people take turns owning an object. Exchanging Kula gifts cements strong friendships, which par-ticipants see as being similar to marriages. Thus Kula gifts bear many of the hallmarks of loved objects: they are beautiful; they help define the owner's identity; and they are quintessential person-thing-person items. It is plausible that many early hu-mans had similar types of possessions, so it is quite likely that they loved them, too.

Although early humans probably loved things from time to time, there are two reasons for believing it didn't happen nearly as often as it does now. First, the early human world was not as saturated with things as ours is. I know five-year-olds today who own far more stuff than a typical aboriginal tribes-person would own during his or her entire lifetime. Second, in hunter-gatherer societies, objects such as Kula gifts are rare. A cross-cultural study conducted by business professors Melanie Wallendorf and Eric Arnould[34] compared the favorite posses-sions of Americans to those of the inhabitants of remote African villages. The Americans' favorites usually helped their owners establish a desired identity and mark important social relation-ships. The African villagers, on the other hand, liked their fa-vorite objects for practical reasons — they worked particularly well or could be sold for a high price if need be, for example. But the Africans didn't imbue the objects with the same kinds of symbolic meanings about their identity that the Americans did. Part of this cross-cultural difference came from the fact that the Americans lived in a big society where they interacted with a lot of people who didn't know them well, so they used their possessions to signal who they were. By contrast, the Africans in this study lived in small, remote villages where everybody knew everyone else, so they didn't need to use their possessions to communicate their identity.

As civilization developed and economic output increased, possessions started to play a growing role in human life. Along with the increased prevalence and variety of possessions, we encounter clear references to loving things, such as the biblical condemnation of the love of money and Plutarch's essay "On Love of Wealth." Fast-forward to the Renaissance, and we see unequivocal cases of the sort of love this book is about. For example, Shakespeare writes that Desdemona "so loves" a handkerchief she received as a gift from Othello that she always has it with her "to kiss and talk to." There it is: full-blown love of an object, including anthropomorphism (she talks to it) and love derived from a person-thing-person connection to Othello.

LOVING THINGS: IT'S NOT A BUG; IT'S A FEATURE

A person's behavior is "evolutionarily optimal" if it results in the creation of the greatest possible number of people who share that person's genes. Loving people is often evolutionarily optimal, but this is not true for loving things. Loving people who share our genes (such as our children and siblings) makes evolutionary sense because it leads us to help them and hence help our genes survive. But the things we love never share our genes, so making sacrifices for them doesn't propagate our genes.

What about loving people who don't share our genes, such as our mates and friends? Mutual love for these people also makes evolutionary sense, but for a different reason. A shared love between friends creates a mutual-aid pact: you help them, and they help you. But this only works when the love is mutual. If loving a friend meant that you helped that person but his or her behavior toward you didn't change, then loving friends would be an evolutionary disadvantage. That's why research[35] finds that one of the most common things that lead us to love other people is realizing that the other people love us back. By contrast, loving things changes our behavior toward them but never changes their behavior toward us. If you type a command into a computer, it doesn't matter whether you love the computer, hate the computer, or couldn't care less about the computer; it will only respond to the exact keystrokes you have entered. So regardless of whether we're considering the evolutionary logic of loving people who share our genes or loving people who don't share our genes, in either case that logic doesn't apply to loving things.

I should clarify that I'm not saying it never makes evolutionary sense to place a high value on something. People will have the best chance of spreading their genes if they value objects exactly in accordance with the item's practical ability to help them survive, attract a mate, have children, and so on. If an object provides a lot of these benefits, we should value it highly and take good care of it. But we should only value things to the extent that they provide these practical benefits, no more and no less.

By contrast, *at its very core, love is about caring for people and things in ways that exceed the practical benefits we get from them.* We make sacrifices for our children that vastly exceed any direct practical benefits we get in exchange. And should the need arise, we are also willing to make sacrifices for other people we love. Similarly, we often invest a huge amount of time and energy in the things we love that is only paid back to us in emotional rewards, not in the kind of practical benefits that improve our evolutionary success. For example, psychologists Jesse Chandler and Norbert Schwarz found[36] that when consumers are deciding if they want to repurchase an anthropomorphic product, their decision is based more on how much they like the product's "personality" than it is on the product's practical benefits. Our tendency to care for the things we love in ways that exceed their practical utility is also evident every time we can't bear to part with something we love but don't use anymore.

Loving things can be seen as a bug in the evolutionary system. But I say, "It's not a bug; it's a feature." Many of the best things in life aren't evolutionarily optimal in that they don't lead us to have more children or otherwise replicate our genes. In fact, most people in the postindustrial world intentionally limit the number of children they have, at least in part so that they have time to pursue the activities they love. The evolutionary story of how love developed can give us a lot of insight into why we think and behave in certain ways. But we'd lead horribly impoverished lives if we never went beyond maximizing our contribution to the gene pool.

10

The Future of the Things We Love

*With the rise of self-driving vehicles, it's only
a matter of time until there's a country song
where the guy's truck leaves him.*

— ANONYMOUS

I N CHAPTER 9, I LOOKED BACK TO EVOLUTIONARY HISTORY AND
its role in shaping the ways people love things. In this final
chapter, I'm going to look forward by exploring three types of
technology: brain-computer interfaces, conversation generators, and consensual telepathy. Each of these technologies will
be embedded in new products that people are likely to love even
more than Apple fans now love their iPhones. And each has
the potential to fundamentally change the experience of being
human. These technologies once struck me as far-fetched, but
there are now working prototypes for all of them.

Throughout the book I have discussed three relationship
warmers — anthropomorphism, people connectors, and integration
into the self — that allow us to think about things in ways that

go beyond a concern with their practical benefits. Each of these future technologies will increase the power of one of the relationship warmers. Brain-computer interfaces will be used in products, and this will dramatically strengthen the feeling that those products are part of ourselves. Conversation generators will allow us to create anthropomorphic machines that people will love with an intensity normally reserved for other people. The third technology, consensual telepathy, will revolutionize the way the things we love connect us to other people. Because these technologies will vastly increase the power of the three relationship warmers, many people will come to love the devices that use these technologies and will do so more intensely than they've loved things in the past.

INTEGRATION INTO THE SELF THROUGH BRAIN-COMPUTER INTERFACES

Consider these quotations from people on Reddit confessing to problems they've had finding their phones:

> More than once I have searched for my phone with the light on my phone, and once while I was talking to my mom on speaker.

> One time I was watching a movie on my phone and went to pull out my phone to check the time. Scariest 5 minutes of my life.

> I have literally been on the phone unable to remember where my phone is.

What's going on? A useful insight, attributed to Bill Gates, states that the "advance of technology is based on making it fit in so that you don't really even notice it, so it's part of everyday life." We see this most clearly in smartphones. But phones aren't just part of their users' lives; they have also become part of the users themselves. I once heard Jerry Seinfeld quip, "They call it an iPhone because it's half you and half your phone." This is why smartphone owners can lose conscious awareness of their phones even when they are in their hands. The object has merged with the owner. Soon enough, technology is going to take this phenomenon to a whole new level.

In 2014, Juliano Pinto made what was both the least impressive and most impressive ceremonial opening kick in the history of soccer's World Cup.[1] The kick itself was weak, rolling around ten feet before being picked up by a referee. More impressive, however, was the fact that Pinto made the kick despite being completely paralyzed from the chest down. He moved using a mechanical exoskeleton attached to his torso and limbs that he controlled with his mind. He wore an object that looked like a bicycle helmet, which contained sensors measuring his brain activity. A computer then used the information from his brain to control the exoskeleton, which moved his legs so that he could stand and kick the ball. This type of system of sensors that detects brain activity and uses it to control a computer is called a "brain-computer interface."

As I noted in chapter 5, when we use an object and feel we have control over its movement, we come to see it as part of our bodies and thus part of our selves. For example, psychologist Ambra Sposito and her colleagues[2] asked one group of research participants to use a long (sixty-centimeter) stick to complete a task that involved moving small objects. She asked another group to complete the same task using a short (twenty-centimeter) stick. After the participants had completed the task, the researchers asked them to estimate how long their forearms were. Participants who used the long stick perceived their forearms as longer than did participants who used the short stick, showing that participants had integrated the tools into their mental representations of their own bodies.

But this subjective sense that a tool is part of one's body is a matter of degree. For, say, a good tennis player, a racket may feel more like part of the body than most other objects do, but it still feels less like part of the body than the hands do. This is partly because tennis players' control over their rackets is not as complete as their control over their hands. It is also because they get direct sensory feedback from their hands, but they only get limited and indirect sensory feedback from their rackets. As brain-computer interface technology advances, allowing us to control objects with our minds and receive direct sensory feedback from those objects, our relationships to products that use a brain-computer interface will become indistinguishable from our relationships to our bodies. That's why Russell Belk has called this type of technology the "prosthetic self."[3] The things most people absolutely, positively identify as "themselves" are their bodies. So when machines attain complete integration with the body, we will fully integrate them into our sense of self and therefore love them far more deeply than we do today.

Types of Brain-Computer Interfaces

For $80 you can buy a toy that will train you to use your brain waves to levitate a Star Wars X-wing fighter (the kind Luke

Skywalker flew). Well, almost — but the brain waves part is true. The toy, called Force Trainer II: Hologram Experience, includes a headset that really does read your brain waves using EEG (electroencephalograph) technology. The more you concentrate, the higher you appear to lift a holographic image of a spaceship. Amazing as this is, the information from your brain that this headset can read is only able to estimate your overall level of concentration. A better consumer-oriented brain sensor, from a company called Emotiv, sells for $300 and allows the user to mentally trigger any four computer commands that the user has trained the device to initiate. This is enough to move a cursor around a computer screen but not much else. To do more, we're going to need much better sensors.

Currently, one of the best available sensors is an fMRI* machine, such as the one that Marcel Just and Tom Mitchell[4] at Carnegie Mellon University use as a "thought recognition" device. These researchers can hand you pictures of various objects, such as a building or a bicycle, and you can look at them in any order while an fMRI scans your brain activity. The computer reading the output of the scan will then tell you in which order you looked at the pictures. That's really cool, but not very practical: fMRI machines are enormous, noisy, and cost up to a million dollars. And even so, the results they produce are still not nearly detailed enough to reach the full potential of brain-computer interface technology.

Another type of sensor is a very small electrode that is surgically implanted in the brain, or between the brain and the skull.

---

* Functional magnetic resonance imaging (fMRI) measures blood flow in the brain, which can tell us which areas of the brain are most active.

In one current version of this technology,[5] sensors are implanted in a participant's brain near the area used for handwriting. The participant then imagines writing something, and a computer "reads" the person's imagined handwritten message. Another approach is to put the sensors over the speech center in the brain and train a computer to recognize the brain impulses created by silently "speaking" words in one's mind.[6]

What Might a Brain-Computer Interface Be Used to Control?

The exoskeleton used by Juliano Pinto was an extension of his body. Brain-computer interfaces are being developed that will allow people to control a car, at which point the car will become, in effect, an enormous and very powerful prosthetic limb. A creative application of prosthetics that works with today's technology comes in the form of a prosthetic Third Thumb, designed by Dani Clode,[7] which makes the hand much more effective at grasping and lifting (see page 267). Although not controlled through a brain-computer interface, the device does change how the user's hand is represented in the brain. Professor Tamar Makin of University College London, who led this research, explained that "to extend our abilities in new and unexpected ways, the brain will need to adapt the representation of the biological body." The current version of the Third Thumb is operated by pressure sensors located under the user's big toes, which is surprisingly easy for people to learn. And users say that "the robotic thumb felt like part of their own body."

Along with extending our bodies, brain-computer interface technology can also extend our mental capacity. For example, what we see and hear could be stored in a database, giving us,

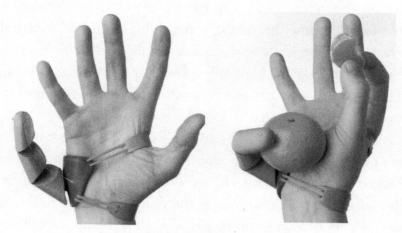

Third Thumb by Dani Clode

literally, photographic memories. We already have access to an astonishing amount of information online, but future technology could make searching the internet feel like searching our own memories.

Will There Be Consumer Demand for Products Using Brain-Computer Interfaces?

Imagine a brain-interface personal computer that you control through mental commands and that responds with visual images and sounds created directly in your brain. Is such a product likely to be a big commercial success? First, the best predictor of a new product's success is not how well it is marketed but rather how genuinely useful it is. And this brain-interface computer could be very useful. Second, assuming it works, users will come to consider it as much a part of themselves as any biological limb or organ. This intense self-integration will lead to an intense love for the product.

On the other hand, the greatest obstacle to this brain-interface computer becoming a megahit is the need for neural sensors to be surgically implanted under the skull. Will this prevent products like the brain-interface computer from ever being viable? Or will it simply slow down their inevitable commercial success?

When a technological innovation is introduced, consumers often say they won't use it because it seems weird, gross, or even dangerous. But opinions change once people get used to it. One of my favorite examples is the Sony Walkman. Before it was introduced, some consumers were shown Walkman prototypes. These consumers said they would never buy one because it would be unbearably weird to walk around in public with headphones on. As a result of these negative comments, Sony delayed releasing the product for many years. Once it was released, though, enough people tried one that it became normal to see people wearing headphones in public. As an increasing number of consumers saw other people wearing headphones, the behavior stopped seeming weird. The social taboo about wearing headphones was broken, and the Walkman became one of the bestselling products in history. Now a company called Beats does a brisk business selling conspicuously large and visually distinctive headphones that are meant to be noticed.

I realize that implanting sensors in the brain is riskier than wearing headphones in public, but even perceived danger won't necessarily prevent people from buying a product if it is otherwise attractive. In the early 1900s, when cars were new, they were far more dangerous than they are today. Early cars had poor brakes, poor handling, and no seat belts. They were driven on unpaved roads, often without posted speed limits, by people without driver's licenses (it wasn't until 1959 that all US states required driving tests). Despite legitimate safety concerns, cars became popular once people experienced their benefits and got used to seeing them around. Early resistance often melts away once a critical mass of consumers is persuaded to adopt the new technology.

Finally, brain-computer interface technology brings us back to an issue covered earlier in this book. When I first introduced the idea that the things we love become part of our selves, I explained that the self has two parts: identity and consciousness. I further explained that — for now, at least — when something we love becomes part of our selves, we include it in our sense of identity, but it does not become part of our consciousness. Brain-computer interface technology is on the verge of changing that. When we reach the point of being able to store and retrieve movies of our past experiences, the devices that allow us to do that will genuinely have become part of our consciousness. And since the depth of our love for things is linked to how strongly they have become part of our selves, our emotional connection to these devices is likely to be intense.

ANTHROPOMORPHISM AND CONVERSATION GENERATORS

Research conducted by psychologist Richard Passman[8] has shown the very important role that beloved blankets and teddy bears play in the lives of small children. But this research also finds that given the choice, kids would rather have their moms. In the contest for kids' affections, the best a teddy bear can hope to win is a silver medal. Along these same lines, studies[9] that compare our love for people with our love for things find that we love people more.

That may change. Currently, our love for things comes closest to our love for people when an object has humanlike characteristics that prompt anthropomorphic thinking. Using today's rather feeble anthropomorphic technology, research[10] has shown that providing elderly people in long-term care facilities with robotic dogs conferred around the same level of therapeutic benefit as giving them real dogs (although this benefit may be short-lived; people have been known to lose interest in robot companions over time).[11] Or consider a virtual interviewer by the name of Ellie that is being used to diagnose soldiers with post-traumatic stress disorder. Ellie asks people questions but doesn't analyze what they say so much as how they say it, looking at their facial expressions, microexpressions, and tones of voice. Studies[12] reviewing Ellie's abilities to diagnose PTSD find that she does around as well as human psychologists, and in some cases even better. What's more, when people talk to Ellie, they tend to reveal more about themselves than they do when they talk to a human psychologist.

When it comes to creating an object that people will form an emotionally rewarding relationship with, the most crucial aspect is the "conversation generator," the software that determines what it will say. The robot dog and the software psychologist have produced impressive therapeutic results even though their conversation generators are extremely limited. What will person-object relationships be like when the objects not only talk to us but also say clever and insightful things? Conversation generators are rapidly moving in that direction. The current talking Barbie, called Hello Barbie, is able to recognize what a child says. It also has an enviable memory. According to her developer, "She should always know that you have two moms and that your grandma died . . . that your favorite color is blue, and that you want to be a veterinarian when you grow up."[13] Hello Barbie can also make empathetic-sounding conversation. When a child confides that she feels shy, Hello Barbie might reply, "Feeling shy is nothing to feel bad about. Just remember this — you made friends with me right away." That's a big step forward from the talking Barbies of yore, which would utter a random sentence such as the infamous "Math is hard" when kids pulled a string on the doll's back.

Just as machines are getting better at conversation, our interpersonal relationships are also changing in ways that make it easier for machines to imitate them. Recall Peter Steiner's famous cartoon showing a dog typing on a computer and saying to the dog next to him, "On the Internet, nobody knows you're a dog." Similarly, it's much easier for a computer to seem human when we interact with it online — and we are having a lot of social interactions online these days. For example, Toyota is installing software called the Toyota Friend into some of its cars. This platform will allow people to communicate with their cars via a social network, just as they communicate with people on Facebook. Currently, the cars aren't as empathetic as Hello Barbie, and they stick to topics like which tires need more air. But Toyota's president and CEO, Akio Toyoda, caught my attention when he said, "I hope cars can become friends with their users and customers will see Toyota as a friend. I want a relationship with my car *in the same way we have a relationship with our friends on social networks*"[14] (emphasis added). What makes that ambition plausible is that on social networks like

Facebook, where nobody has a body and all communication is typed, we enter the computer's "home turf."

I have heard people dismiss concerns about future person-thing relationships crowding out person-person relationships by saying that person-person relationships will always be deeper and more richly rewarding. I agree. But I'm still worried. And here, my perspective as a marketing professor comes into play. For example, people eat a lot of junk food. And even back when TV was pretty terrible, we still watched a lot of it. Given the choice between something that's easy but mediocre and something that's difficult but deeply rewarding, easy wins a whole lot of the time. Meaningful relationships with people aren't always easy. And even when they are going well, human relationships always involve give and take — if you want your friends to listen to your boring stories, you need to listen to their boring stories. But anthropomorphic machines will happily make it all about you, all the time. This ability for people to be totally selfish in their relationships with machines not only makes anthropomorphic machines, sadly, attractive, it may also damage our human relationships if we start expecting people to indulge us the same way robot companions do.

PEOPLE CONNECTORS AND CONSENSUAL TELEPATHY

People connectors, which are objects that become meaningful to us because of their central place in person-thing-person relationships, may also be supercharged by new technology. Contrary to the possibility that future anthropomorphic objects might weaken our human relationships, here technology might bring us closer to other people. Moreover, if it works, it will fundamentally extend the limits of what it means to love someone.

What would happen, for example, if brain-computer interface technology was used not to connect your brain to, say, a robotic arm but rather to connect it directly to another person's brain? Is this even possible?

Although it may sound far-fetched, a simple working version of this technology already exists. In an experiment conducted by physician Miguel Nicolelis and his colleagues,[15] a rat in Brazil and a rat in the United States were each placed in a cage with two containers, one of which held a treat while the other was empty. If the rat first opened the container that held the treat, it got the treat. But if it chose the wrong container, it got nothing. Each container had a little light in front of it. For the Brazilian rat, the scientists would turn on the light in front of the container with the treat. So it didn't take the rat long to learn that the treat was always in the container with the light on. The US rat had no such luck. Both lights were always on, so they didn't provide any information. However, if the treat was in, say, the left container in Brazil, it would also be in the left container in the United States. If the Brazilian rat could just tell the American rat which container to look in first, the American rat would also get a treat. But the rats were on different continents and didn't have Zoom, so there was no way for them to communicate . . . or was there?

Here's where things get amazing. Both rats had neural implants in their brains. These implants recorded the brain activity of the Brazilian rat and transmitted it to the US rat, whose implants stimulated the corresponding parts of its brain. At first, the US rat only selected the container that held the treat

50 percent of the time. But gradually, the US rat learned to interpret the brain signals sent by the Brazilian rat and started to select the correct container much more often. The researchers then changed the rewards a bit, so that when the US rat chose the correct container, the Brazilian rat would get an extra treat. Even though the Brazilian rat had no idea the US rat existed, its brain noticed that when it did certain things, it would get an extra treat. With practice, the Brazilian rat's brain started to figure out what these certain things were so it could do them more often, which meant it sent clearer signals to the US rat, and the US rat became even more accurate in its choices. That's one of the things that makes this technology viable: over time, the brains of both the sender and the receiver adapt to increase the effectiveness of the signaling mechanism.

How long will it take to establish brain-to-brain communication among people? Through his company Neuralink, tech entrepreneur Elon Musk is trying to develop brain-to-brain interfacing technology. In his proposed system, we would be able to share a visual image, a sound, or a feeling directly with another person without describing it in words. As Musk told journalist Tim Urban in an interview:[16]

> There are a bunch of concepts in your head that then your brain has to try to compress into this incredibly low data rate called speech or typing. That's what language is — your brain has executed a compression algorithm on thought, on concept transfer. . . . If you have two brain interfaces, you could actually do an uncompressed direct conceptual communication with another person.

In Urban's view, Neuralink eclipses both Tesla's plans to mass-market electric vehicles and SpaceX's ambitions to send humans to Mars. Whereas Tesla and SpaceX "aim to redefine what future humans will do — Neuralink wants to redefine what future humans will be." How long before this science fiction technology, which would change the fundamental nature of human existence, comes online? Musk estimates eight to ten years.

HBO's *Made for Love* puts this technology in a sinister light. While the real Elon Musk refers to it as "consensual telepathy," the tech billionaire in the show seeks to impose a similar technology on his wife against her will. It is meant to lead to complete mental union, with no secrets, no privacy, and no escape. This drives home the point that while the idea of two people becoming one sounds lovely in a wedding ceremony, like all good things, it could be taken too far.

While a direct brain-to-brain interface would be revolutionary, it wouldn't be entirely alien to human experience. Garth Fletcher[17] and other psychologists study mind reading in intimate relationships, by which they mean the heightened ability of people in close relationships to intuit what their partners are thinking and feeling. The brain waves of close friends and romantic partners also tend to be in sync with one another. In one study, UCLA psychologist Carolyn Parkinson and her colleagues[18] recorded the brain waves of forty-two volunteers as they watched a series of short video clips — the typical stuff you'd find on YouTube. The group of volunteers included sets of close friends, friendly acquaintances, and sets of strangers. The findings showed that if two people were friends, their brain waves synced up as they watched the videos, but this was far less common among strangers. The researchers believe that when two people meet and "click" with each other, it may be that they are thinking in sync, and this leads them to become friends. Research conducted by neuroscientist Pavel Goldstein and his colleagues[19] found that when romantic partners empathize with each other, their brain waves also synchronize. This synchronization is further increased when the partners touch each other. In its most benign form, brain-to-brain technology may be, in effect, a kind of "touch 2.0" that fully synchronizes our mental experiences with those of another person. If so, it would be the ultimate example of something that enhances human connection.

CLOSING THOUGHTS

Erich Fromm[20] wrote, "Love is the only sane and satisfactory answer to the problem of human existence." This is certainly true for the people we love. But it also extends to many of the things we love, particularly when they help us construct our identities and connect us to other people.

When we physically surround ourselves with objects we love and spend our time on activities we love, these love objects become part of the world in which we live. Yet the things we love are also part of our selves. Because the things we love are both part of ourselves and part of the world around us, they blur the distinction between the self and the world. In this and other ways, understanding our love of things can help us see that the boundary between ourselves and the rest of the universe is fuzzier and more flexible than it sometimes appears.

As I said in chapter 9, love began in ancient animals as self-concern. Eventually it expanded to include family members, then expanded again to include friends, then expanded yet again to include all sorts of things. It's an inspiring trajectory, in which love reaches out from the self toward ever greater parts of the world. Humans' ability to love such a wide variety of things allows us to bring love into many facets of our lives, not just our social relationships. It is certainly my hope that by understanding ourselves and the things we love a little better, we can build lives that are more enjoyable as well as more connected to the people and the world around us.

Acknowledgments

I MUST BEGIN BY EXTENDING HUGE THANKS TO MY WIFE, AURA, WHO supported me in every possible way through the long process of writing this book. She not only provided encouragement and made it possible for me to have the time I needed for this work, she also served as a discussion partner for ideas and carefully edited the content. I am also grateful for, and to, my children, Isaac and Jonah. At the start of this project they provided me with insights into the ways in which children experience the world, and by the time it was complete, they were sharing their insights into the ideas it contains.

I am deeply grateful to the many people who provided thoughtful feedback on earlier versions of this work, including John Abramyan, Mara Adelman, Jeff and Rena Basch, Hannah Bernard, Ruth Bernard, Corry Buckwalter, Paula Caproni, Christine Chastain, Darren Dahl, Robert Diener, Scott Foster, Ben and Shari Fox, Leah and Ron Gilbert, Beth Greenapple, Dan Haybron, David Isaacson, Mark Jannot, Nomi and Cory

Joyrich, Ann Manikas, Rebekah Modrak, Janice Molloy, Alan and Margaret Ness, Ruth Ness, David Rosenfeld, Matt Roth, Jill Sundie, Neil Thin, Jean Timsit, Jeremy Wood, and Larry J. Young.

I would also like to thankfully recognize my collaborators in the research on which this book is based, including Mara Adelman, Rick Bagozzi, Rajeev Batra, Philipp Rauschnabel, Aric Rindfleisch, and Nancy Wong. In the research projects we conducted together it is impossible to say where one person's ideas stop and another's start. I'm sure that many of the ideas in this book have sprung from the minds of my collaborators. I'm grateful for these contributions.

I have been pleasantly amazed by the contributions of my wonderful editor, Tracy Behar, whose advice was always right — even in those rare instances when I didn't take it. I have also been blown away by the skill, dedication, and sensitivity of copyeditor Barbara Clark and the entire Little, Brown Spark team: Ian Straus, Pat Jalbert-Levine, Jessica Chun, and Juliana Horbachevsky.

Special honors are due to my agent extraordinaire, Esmond Harmsworth. He showed tremendous love for this project by nurturing it for more than a decade, far longer than his genes would have seen as evolutionarily optimal given its expected returns. Through our back-and-forth on seven (!) different versions of the book proposal, the manuscript moved from being a hodgepodge of findings and anecdotes to (what I hope is) a coherent overarching theory of how and why people love things. Throughout this process he continued to provide detailed feedback on both the content and the writing, without which this project would never have seen the light of day.

Selected Bibliography

Ahuvia, Aaron C. "Beyond the Extended Self: Loved Objects and Consumers' Identity Narratives." *Journal of Consumer Research* 32, no. 1 (June 2005): 171–84. https://doi.org/10.1086/429607.

———. "For the Love of Money: Materialism and Product Love." In *Meaning, Measure, and Morality of Materialism,* edited by Floyd W. Rudmin and Marsha Lee Richins. Provo, UT: Association for Consumer Research, 1992.

———. "I Love It! Towards a Unifying Theory of Love Across Diverse Love Objects." (Abridged PhD diss., Northwestern University, 1993). https://deepblue.lib.umich.edu/handle/2027.42/35351.

———. "Nothing Matters More to People Than People: Brand Meaning and Social Relationships." *Review of Marketing Research* 12 (May 2015): 121–49. https://doi.org/10.1108/S1548-643520150000012005. Lead article, special issue on brand meaning.

Ahuvia, Aaron C., et al. "Pride of Ownership: An Identity-Based Model." *Journal of the Association for Consumer Research* 3, no. 2 (April 2018): 1–13. https://doi.org/10.1086/697076.

Ahuvia, Aaron C., Rajeev Batra, and Richard P. Bagozzi. "Love, Desire and Identity: A Conditional Integration Theory of the Love of Things."

In *The Handbook of Brand Relationships*, edited by Deborah J. MacInnis, C. Whan Park, and Joseph R. Priester. New York: M. E. Sharpe, 2009.

Ahuvia, Aaron C., Philipp Rauschnabel, and Aric Rindfleisch. "Is Brand Love Materialistic?" *Journal of Product & Brand Management* 30, no. 3 (December 2020): 467–80. https://doi.org/10.1108/JPBM-09-2019-2566.

Bagozzi, Richard P., Rajeev Batra, and Aaron C. Ahuvia. "Brand Love: Development and Validation of a Practical Scale." *Marketing Letters* 28 (September 2016): 1–14. https://doi.org/10.1007/s11002-016-9406-1. Lead article.

Batra, Rajeev, Aaron C. Ahuvia, and Richard P. Bagozzi. "Brand Love." *Journal of Marketing* 76, no. 2 (March 2012): 1–16. https://doi .org/10.1509/jm.09.0339. Lead article and runner-up for the Harold H. Maynard Award for the best *Journal of Marketing* article on marketing theory. All authors contributed equally to this work.

Carroll, Barbara A., and Aaron C. Ahuvia. "Some Antecedents and Outcomes of Brand Love." *Marketing Letters* 17, no. 2 (April 2006): 79–89. https://doi.org/10.1007/s11002-006-4219-2.

Rauschnabel, Philipp, et al. "The Personality of Brand Lovers: An Examination in Fashion Branding." In *Consumer Brand Relationships: Meaning, Measuring, Managing,* edited by Marc Fetscherin and Tobias Heilmann. London: Palgrave Macmillan, 2015.

Rauschnabel, Philipp A., and Aaron C. Ahuvia. "You're So Lovable: Anthropomorphism and Brand Love." *Journal of Brand Management* 21, no. 5 (August 2014): 372–95. https://doi.org/10.1057/bm.2014.14.

Wong, Nancy, and Aaron C. Ahuvia. "Personal Taste and Family Face: Luxury Consumption in Confucian and Western Societies." *Psychology & Marketing* 15, no. 5 (1998): 423–41. https://doi.org/10.1002/(SICI) 1520-6793(199808)15:5<423::AID-MAR2>3.0.CO;2-9.

Notes

CHAPTER 1: A Many-Splendored Thing

1. A. Guttmann, "Media Spending Worldwide 2014–2022," Statista, August 9, 2019, https://www.statista.com/statistics/273288/advertising-spending-worldwide/.
2. Daniel M. Haybron, "Central Park: Nature, Context, and Human Wellbeing," *International Journal of Wellbeing* 1, no. 2 (2011): 235–54, https://doi.org/10.5502/ijw.v1i2.6.
3. Rajeev Batra, Aaron Ahuvia, and Richard P. Bagozzi, "Brand Love," *Journal of Marketing* 76, no. 2 (2012): 1–16, https://doi.org/10.1509/jm.09.0339; Richard P. Bagozzi, Rajeev Batra, and Aaron Ahuvia, "Brand Love: Development and Validation of a Practical Scale," *Marketing Letters* 28 (2016): 1–14, https://doi.org/10.1007/s11002-016-9406-1.
4. Terence A. Shimp and Thomas J. Madden, "Consumer-Object Relations: A Conceptual Framework Based Analogously on Sternberg's Triangular Theory of Love," *Advances in Consumer Research* 15 (1988): 163–68.
5. Aaron Ahuvia, "I Love It! Towards a Unifying Theory of Love Across Diverse Love Objects," abridged (PhD diss., Northwestern University, 1993), https://deepblue.lib.umich.edu/handle/2027.42/35351.

6. Bernard I. Murstein, "A Taxonomy of Love," in *The Psychology of Love*, ed. Robert Sternberg and Michael L. Barnes (New Haven, CT: Yale University Press, 1988), 13–37.

7. Sandra L. Murray, John G. Holmes, and Dale W. Griffin, "The Benefits of Positive Illusions: Idealization and the Construction of Satisfaction in Close Relationships," *Journal of Personality and Social Psychology* 70, no. 1 (1996): 79–98, https://doi.org/10.1037//0022-3514.70.1.79.

8. Christoph Patrick Werner et al., "Price Information Influences the Subjective Experience of Wine: A Framed Field Experiment," *Food Quality and Preference* 92 (2021): 104223, https://doi.org/10.1016/j.foodqual.2021.104223.

9. Joseph W. Alba and Elanor F. Williams, "Pleasure Principles: A Review of Research on Hedonic Consumption," *Journal of Consumer Psychology* 23, no. 1 (2013): 2–18, https://doi.org/10.1016/j.jcps.2012.07.003.

10. From data collected for Batra, Ahuvia, and Bagozzi, "Brand Love."

11. Zick Rubin, "Measurement of Romantic Love," *Journal of Personality and Social Psychology* 16, no. 2 (1970): 256–73, https://doi.org/10.1037/h0029841.

12. Ahuvia, "I Love It!"

13. Ahuvia, unpublished data.

14. From data collected for Batra, Ahuvia, and Bagozzi, "Brand Love."

15. Aaron C. Ahuvia, Rajeev Batra, and Richard P. Bagozzi, "Love, Desire and Identity: A Conditional Integration Theory of the Love of Things," in *The Handbook of Brand Relationships*, ed. Deborah J. MacInnis, C. Whan Park, and Joseph R. Priester (New York: M. E. Sharpe, 2009).

16. Ahuvia, Batra, and Bagozzi, "Love, Desire and Identity."

17. "'Nones' on the Rise," Pew Research Center, October 9, 2012, https://www.pewforum.org/2012/10/09/nones-on-the-rise/.

18. "'Nones' on the Rise."

19. Russell Belk and Gülnur Tumbat, "The Cult of Macintosh," *Consumption Markets and Culture* 8, no. 3 (2005): 205–17, https://doi.org/10.1080/10253860500160403.

20. Jonah Weiner, "Jerry Seinfeld Intends to Die Standing Up," *New York Times*, December 20, 2012.

21. Ron Shachar et al., "Brands: The Opiate of the Nonreligious Masses?," *Marketing Science* 30, no. 1 (2011): 92–110, https://doi.org/10.1287/mksc.1100.0591.

22. Batra, Ahuvia, and Bagozzi, "Brand Love"; Bagozzi, Batra, and Ahuvia, "Brand Love."

23. Wendy Maxian et al., "Brand Love Is in the Heart: Physiological Responding to Advertised Brands," *Psychology & Marketing* 30, no. 6 (2013): 469–78, https://doi.org/10.1002/mar.20620.

24. Ahuvia, "I Love It!"

25. Ahuvia, "I Love It!"

26. Ahuvia, "I Love It!"

27. Vanitha Swaminathan, Karen M. Stilley, and Rohini Ahluwalia, "When Brand Personality Matters: The Moderating Role of Attachment Styles," *Journal of Consumer Research* 35, no. 6 (2009): 985–1002, https://doi.org/10.1086/593948.

28. Matthew Thomson, Jodie Whelan, and Allison R. Johnson, "Why Brands Should Fear Fearful Consumers: How Attachment Style Predicts Retaliation," *Journal of Consumer Psychology* 22, no. 2 (2012): 289–98, https://doi.org/10.1016/j.jcps.2011.04.006.

29. John L. Lastovicka and Nancy J. Sirianni, "Truly, Madly, Deeply: Consumers in the Throes of Material Possession Love," *Journal of Consumer Research* 38, no. 2 (2011): 323–42, https://doi.org/10.1086/658338.

30. Ahuvia, "I Love It!"

31. Ahuvia, "I Love It!"

32. Carolyn Yoon et al., "A Functional Magnetic Resonance Imaging Study of Neural Dissociations Between Brand and Person Judgments," *Journal of Consumer Research* 33, no. 1 (2006): 31–40, https://doi.org/10.1086/504132.

33. Iskra Herak, Nicolas Kervyn, and Matthew Thomson, "Pairing People with Products: Anthropomorphizing the Object, Dehumanizing the Person," *Journal of Consumer Psychology* 30, no. 1 (2020), 125–39, https://doi.org/10.1002/jcpy.1128.

34. Martha Nussbaum, "Objectification," *Philosophy & Public Affairs* 24, no. 4 (1995): 251–54.

35. Aaron C. Ahuvia, "Beyond the Extended Self: Loved Objects and Consumers' Identity Narratives," *Journal of Consumer Research* 32, no. 1 (2005): 171–84, https://doi.org/10.1086/429607.

36. Lasana T. Harris and Susan T. Fiske, "Dehumanized Perception: The Social Neuroscience of Thinking (or Not Thinking) About Disgusting People," in *European Review of Social Psychology* vol. 20, ed.

Wolfgang Stroebe and Miles Hewstone (London: Psychology Press, 2010), 192–231.

37. Andreas Fürst et al., "The Neuropeptide Oxytocin Modulates Consumer Brand Relationships," *Scientific Reports* 5 (2015): 14960, https://doi.org/10.1038/srep14960.

38. Martin Reimann et al., "How We Relate to Brands: Psychological and Neurophysiological Insights into Consumer-Brand Relationships," *Journal of Consumer Psychology* 22, no. 1 (2012): 128–42, https://doi.org/10.1016/j.jcps.2011.11.003.

39. Martin Reimann, Sandra Nuñez, and Raquel Castaño, "Brand-Aid," *Journal of Consumer Research* 44, no. 3 (2017): 673–91, https://doi.org/10.1093/jcr/ucx058.

CHAPTER 2: **Honorary People**

1. See Holger Luczak et al., "PALAVER: Talking to Technical Devices," in *Proceedings of the International Conference on Affective Human Factors Design*, ed. Martin G. Helander, Halimahtun M. Khalid, and Ming Po Tham (London: ASEAN Academic Press, 2001), 349–55.

2. "Progressive.com Surveys Americans to Determine How Much We Love Our Cars," Auto Channel, February 7, 2001, https://www.theautochannel.com/news/2001/02/07/014192.html.

3. Christoph Bartneck et al., "The Influence of Robot Anthropomorphism on the Feelings of Embarrassment When Interacting with Robots," *Paladyn: Journal of Behavioral Robotics* 1, no. 2 (2010): 109–15, https://doi.org/10.2478/s13230-010-0011-3.

4. Sara Kim and Ann L. McGill, "Gaming with Mr. Slot or Gaming the Slot Machine? Power, Anthropomorphism, and Risk Perception," *Journal of Consumer Research* 38, no. 1 (2011): 94–107, https://doi.org/10.1086/658148.

5. This is the first of several images depicting materials used in psychological studies. Many of the original images would not reproduce clearly in this book, so I decided to have them redrawn. The originals can readily be found in the published papers cited in these endnotes.

6. Valeria Gazzola et al., "The Anthropomorphic Brain: The Mirror Neuron System Responds to Human and Robotic Actions," *NeuroImage* 35, no. 4 (2007): 1674–84, https://doi.org/10.1016/j.neuroimage.2007.02.003.

7. Lasana T. Harris and Susan T. Fiske, "The Brooms in *Fantasia*: Neural Correlates of Anthropomorphizing Objects," *Social Cognition* 26, no. 2 (2008): 210–23, https://doi.org/10.1521/soco.2008.26.2.210.

8. Sonja Windhager et al., "Face to Face: The Perception of Automotive Designs," *Human Nature* 19, no. 4 (2008): 331–46, https://doi.org/10.1007/s12110-008-9047-z.

9. Jan R. Landwehr, Ann L. McGill, and Andreas Herrmann, "It's Got the Look: The Effect of Friendly and Aggressive 'Facial' Expressions on Product Liking and Sales," *Journal of Marketing* 75, no. 3 (2011): 132–46, https://doi.org/10.1509/jmkg.75.3.132.

10. Maferima Touré-Tillery and Ann L. McGill, "Who or What to Believe: Trust and the Differential Persuasiveness of Human and Anthropomorphized Messengers," *Journal of Marketing* 79, no. 4 (2015): 94–110, https://doi.org/10.1509/jm.12.0166.

11. Holger Luczak, Matthias Roetting, and Ludger Schmidt, "Let's Talk: Anthropomorphization as Means to Cope with Stress of Interacting with Technical Devices," *Ergonomics* 46, no. 13–14 (2003): 1361–74, https://doi.org/10.1080/00140130310001610883.

12. Andrew Ortony, Gerald L. Clore, and Allan Collins, *The Cognitive Structure of Emotions* (Cambridge, UK: Cambridge University Press, 1988).

13. Hilary Downey and Sarah Ellis, "Tails of Animal Attraction: Incorporating the Feline into the Family," *Journal of Business Research* 61, no. 5 (2008): 434–41, https://doi.org/10.1016/j.jbusres.2007.07.015; S. Shyam Sundar, "Loyalty to Computer Terminals: Is It Anthropomorphism or Consistency?," *Behaviour & Information Technology* 23, no. 2 (2004): 107–18, https://doi.org/10.1080/0144929031000165 9222.

14. Phillip M. Hart, Sean R. Jones, and Marla B. Royne, "The Human Lens: How Anthropomorphic Reasoning Varies by Product Complexity and Enhances Personal Value," *Journal of Marketing Management* 29, no. 1–2 (2013): 105–21, https://doi.org/10.1080/0267257X.2012.759993.

15. Jesse Chandler and Norbert Schwarz, "Use Does Not Wear Ragged the Fabric of Friendship: Thinking of Objects as Alive Makes People Less Willing to Replace Them," *Journal of Consumer Psychology* 20, no. 2 (2010): 138–45, https://doi.org/10.1016/j.jcps.2009.12.008.

16. Philipp A. Rauschnabel and Aaron C. Ahuvia, "You're So Lovable: Anthropomorphism and Brand Love," *Journal of Brand Management* 21, no. 5 (August 2014): 372–95, https://doi.org/10.1057/bm .2014.14.

17. Rauschnabel and Ahuvia, "You're So Lovable"; Deborah J. MacInnis and Valerie S. Folkes, "Humanizing Brands: When Brands Seem to Be Like Me, Part of Me, and in a Relationship with Me," *Journal of Consumer Psychology* 27, no 3 (2017): 355–74, https://doi.org/10.1016/j .jcps.2016.12.003.

18. Adam Waytz, Joy Heafner, and Nicholas Epley, "The Mind in the Machine: Anthropomorphism Increases Trust in an Autonomous Vehicle," *Journal of Experimental Social Psychology* 52 (2014): 113–17, https://doi.org/10.1016/j.jesp.2014.01.005.

19. Sara Kim, Rocky Peng Chen, and Ke Zhang, "Anthropomorphized Helpers Undermine Autonomy and Enjoyment in Computer Games," *Journal of Consumer Research* 43, no. 2 (2016): 282–302, https://doi.org/10.1093/jcr/ucw016.

20. Jing Wan and Pankaj Aggarwal, "Befriending Mr. Clean: The Role of Anthropomorphism in Consumer-Brand Relationships," in *Strong Brands, Strong Relationships*, ed. Susan Fournier, Michael J. Breazeale, and Jill Avery (Abingdon, UK: Routledge, 2015), 119–34.

21. Pankaj Aggarwal and Ann L. McGill, "Is That Car Smiling at Me? Schema Congruity as a Basis for Evaluating Anthropomorphized Products," *Journal of Consumer Research* 34, no. 4 (2007): 468–79, https://doi.org/10.1086/518544.

22. Simon Hudson et al., "The Influence of Social Media Interactions on Consumer-Brand Relationships: A Three-Country Study of Brand Perceptions and Marketing Behaviors," *International Journal of Research in Marketing* 33, no. 1 (2016): 27–41, https://doi.org/10.1016/j .ijresmar.2015.06.004.

23. Hyokjin Kwak, Marina Puzakova, and Joseph Rocereto, "Better Not Smile at the Price: The Differential Role of Brand Anthropomorphization on Perceived Price Fairness," *Journal of Marketing* 79, no. 4 (2015): 56–76, https://doi.org/10.1509/jm.13.0410.

24. Marina Puzakova, Hyokjin Kwak, and Joseph Rocereto, "When Humanizing Brands Goes Wrong: The Detrimental Role of Brand Anthropomorphization Amidst Product Wrongdoings," *Journal of Marketing* 77, no. 3 (2013): 81–100.

25. Luczak, Roetting, and Schmidt, "Let's Talk."

26. Kate Letheren et al., "Individual Difference Factors Related to Anthropomorphic Tendency," *European Journal of Marketing* 50, no. 5–6 (2016): 973–1002, https://doi.org/10.1108/EJM-05-2014-0291.

27. Shankar Vedantam, "The Lonely American Man," October 14, 2019, *Hidden Brain,* produced by Tara Boyle, podcast, 16:34, https://pca.st/episode/001413bf-71ad-42ca-a445-67129e144ffc?t=994.0.

28. Nicholas Epley et al., "When We Need a Human: Motivational Determinants of Anthropomorphism," *Social Cognition* 26, no. 2 (2008): 143–55, https://doi.org/10.1521/soco.2008.26.2.143.

29. Friederike Eyssel and Natalia Reich, "Loneliness Makes the Heart Grow Fonder (of Robots): On the Effects of Loneliness on Psychological Anthropomorphism," *2013 8th ACM/IEEE International Conference on Human-Robot Interaction (HRI)* (2013): 121–22, https://doi.org/10.1109/HRI.2013.6483531.

30. Adam Waytz et al., "Making Sense by Making Sentient: Effectance Motivation Increases Anthropomorphism," *Journal of Personality and Social Psychology* 99, no. 3 (2010): 410–35, https://doi.org/10.1037/a0020240.

31. Luczak, Roetting, and Schmidt, "Let's Talk."

32. Mary M. Herrald, Joe Tomaka, and Amanda Y. Medina, "Pet Ownership Predicts Adherence to Cardiovascular Rehabilitation," *Journal of Applied Social Psychology* 32, no. 6 (2002): 1107–23, https://doi.org/10.1111/j.1559-1816.2002.tb01428.x.

33. Tori Rodriguez, "Pets Help Us Achieve Goals and Reduce Stress," *Scientific American,* November 1, 2012.

34. Letheren et al., "Individual Difference Factors Related to Anthropomorphic Tendency."

35. Paul M. Connell, "The Role of Baseline Physical Similarity to Humans in Consumer Responses to Anthropomorphic Animal Images," *Psychology & Marketing* 30, no. 6 (2013): 461–68, https://doi.org/10.1002/mar.20619.

36. Aaron Ahuvia, "Commentary on Exploring the Dark Side of Pet Ownership: Status- and Control-Based Pet Consumption: A Reinterpretation of the Data," *Journal of Business Research* 61, no. 5 (2008): 497–99, https://doi.org/10.1016/j.jbusres.2007.01.028.

37. Michael B. Beverland, Francis Farrelly, and Elison Ai Ching Lim, "Exploring the Dark Side of Pet Ownership: Status- and Control-Based Pet Consumption," *Journal of Business Research* 61, no. 5 (2008): 490–96, https://doi.org/10.1016/j.jbusres.2006.08.009.

38. Reported in Daisy Yuhas, "Pets: Why Do We Have Them?," *Scientific American Mind* 26, no. 3 (2015): 28–33, https://doi.org/10.1038/scientificamericanmind0515-28.

39. Stephen Kellett et al., "Compulsive Hoarding: An Interpretative Phenomenological Analysis," *Behavioural and Cognitive Psychotherapy* 38, no. 2 (2010): 141–55, https://doi.org/10.1017/S1352465809990622.

40. Kiara R. Timpano and Ashley M. Shaw, "Conferring Humanness: The Role of Anthropomorphism in Hoarding," *Personality and Individual Differences* 54, no. 3 (2014): 383–88, https://doi.org/10.1016/j.paid.2012.10.007.

41. Melissa M. Norberg et al., "Anxious Attachment and Excessive Acquisition: The Mediating Roles of Anthropomorphism and Distress Intolerance," *Journal of Behavioral Addictions* 7, no. 1 (2018): 171–80, https://doi.org/10.1556/2006.7.2018.08.

42. James Vlahos, "Barbie Wants to Get to Know Your Child," *New York Times Magazine*, September 16, 2015.

43. See Adam Waytz, John Cacioppo, and Nicholas Epley, "Who Sees Human? The Stability and Importance of Individual Differences in Anthropomorphism," *Perspectives on Psychological Science* 5, no. 3 (2010): 219–32, https://doi.org/10.1177/1745691610369336; also see Adam Waytz, Nicholas Epley, and John Cacioppo, "Social Cognition Unbound: Insights into Anthropomorphism and Dehumanization," *Current Directions in Psychological Science* 19, no. 1 (2010): 58–62, https://doi.org/10.1177/0963721409359302.

44. Mark Levine, "Share My Ride," *New York Times Magazine*, March 5, 2009.

45. Hee-Kyung Ahn, Hae Joo Kim, and Pankaj Aggarwal, "Helping Fellow Beings: Anthropomorphized Social Causes and the Role of Anticipatory Guilt," *Psychological Science* 25, no. 1 (2014): 224–29, https://doi.org/10.1177/0956797613496823.

46. Kellett et al., "Compulsive Hoarding."

47. Chandler and Schwarz, "Use Does Not Wear Ragged the Fabric of Friendship."

CHAPTER 3: What Does It Mean to Have a Relationship with a Thing?

1. Helen Fisher, Arthur Aron, and Lucy L. Brown, "Romantic Love: An fMRI Study of a Neural Mechanism for Mate Choice," *Journal*

of Comparative Neurology 493, no. 1 (2005): 58–62, https://doi.org /10.1002/cne.20772.

2. Sarah Broadbent, "Brand Love in Sport: Antecedents and Consequences" (PhD diss., School of Management and Marketing, Deakin University, 2012), https://dro.deakin.edu.au/view/DU:30062512.

3. Elaine Hatfield and Richard Rapson, "Love and Attachment Processes," in *Handbook of Emotions*, ed. Michael Lewis and Jeannette M. Haviland (New York: Guilford Publications, 1993); Marsha L. Richins, "Measuring Emotions in the Consumption Experience," *Journal of Consumer Research* 24, no. 2 (1997): 127–46; Lisa A. Cavanaugh, James R. Bettman, and Mary Frances Luce, "Feeling Love and Doing More for Distant Others: Specific Positive Emotions Differentially Affect Prosocial Consumption," *Journal of Marketing Research* 52, no. 5 (2015): 657–73, https://doi.org/10.1509/jmr.10.0219; Fleur J. M. Laros and Jan-Benedict E. M. Steenkamp, "Emotions in Consumer Behavior: A Hierarchical Approach," *Journal of Business Research* 58, no. 10 (2005): 1437–45, https://doi.org/10.1016/j .jbusres.2003.09.013; Phillip R. Shaver, Shelley Wu, and Judith C. Schwartz, "Cross-Cultural Similarities and Differences in Emotion and Its Representation: A Prototype Approach," in *Emotion: Review of Personality and Social Psychology* 13, ed. Margaret S. Clark (Newbury Park, CA: Sage Publications, 1992).

4. Makenzie J. O'Neil et al., "Prototype Facial Response to Cute Stimuli: Expression and Recognition" (unpublished manuscript, 2019).

5. For more about consumer-brand relationships, see Susan Fournier, "Consumers and Their Brands: Developing Relationship Theory in Consumer Research," *Journal of Consumer Research* 24, no. 4 (March 1998): 343–73; Susan Fournier, "Lessons Learned About Consumers' Relationships with Their Brands," in *The Handbook of Brand Relationships*, ed. Deborah J. MacInnis, C. Whan Park, and Joseph R. Priester (New York: M. E. Sharpe, 2009); Jennifer Aaker, Susan Fournier, and S. Adam Brasel, "When Good Brands Do Bad," *Journal of Consumer Research* 31, no. 1 (June 2004): 1–16; Susan Fournier and Julie L. Yao, "Reviving Brand Loyalty: A Reconceptualization Within the Framework of Consumer-Brand Relationships," *International Journal of Research in Marketing* 14, no. 5 (December 1997): 451–72.

6. Matthew Thomson and Allison R. Johnson, "Marketplace and Personal Space: Investigating the Differential Effects of Attachment Style

Across Relationship Contexts," *Psychology & Marketing* 23, no. 8 (2006): 711–26, https://doi.org/10.1002/mar.20125.

7. Julie Fitness and Garth J. O. Fletcher, "Love, Hate, Anger, and Jealousy in Close Relationships: A Prototype and Cognitive Appraisal Analysis," *Journal of Personality and Social Psychology* 65, no. 5 (1993): 942–58, https://doi.org/10.1037/0022-3514.65.5.942.

8. Robert J. Sternberg, "Explorations of Love," in *Advances in Personal Relationships* vol.1, ed. Warren H. Jones and Daniel Perlman (Greenwich, CT: JAI Press, 1987).

9. Youngme Moon, "Intimate Exchanges: Using Computers to Elicit Self-Disclosure from Consumers," *Journal of Consumer Research* 26, no. 4 (2000): 323–39, https://doi.org/10.1086/209566.

10. John M. Gottman, *Marital Interaction: Experimental Investigations* (New York: Academic Press, 1979).

11. Carol Werner and Bibb Latane, "Interaction Motivates Attraction: Rats Are Fond of Fondling," *Journal of Personality and Social Psychology* 29, no. 3: 328–34, https://doi.org/10.1037/h0035976.

12. Aaron C. Ahuvia, Rajeev Batra, and Richard P. Bagozzi, "Love, Desire and Identity: A Conditional Integration Theory of the Love of Things," in *The Handbook of Brand Relationships*, ed. Deborah J. MacInnis, C. Whan Park, and Joseph R. Priester (New York: M. E. Sharpe, 2009).

13. Alokparna Basu Monga, "Brand as a Relationship Partner: Gender Differences in Perspectives," *Advances in Consumer Research* 29, no. 1 (2002): 36–41.

14. Jodie Whelan et al., "Relational Domain Switching: Interpersonal Insecurity Predicts the Strength and Number of Marketplace Relationships," *Psychology & Marketing* 33, no. 6 (2016): 465–79, https://doi.org/10.1002/mar.20891.

15. Aaron Ahuvia, "I Love It! Towards a Unifying Theory of Love Across Diverse Love Objects," abridged (PhD diss., Northwestern University, 1993), https://deepblue.lib.umich.edu/handle/2027.42/35351.

16. Mark S. Rosenbaum et al., "A Cup of Coffee with a Dash of Love: An Investigation of Commercial Social Support and Third-Place Attachment," *Journal of Service Research* 10, no. 1 (2007): 43–58, https://doi.org/10.1177/1094670507303011.

17. Morgan K. Ward and Darren W. Dahl, "Should the Devil Sell Prada? Retail Rejection Increases Aspiring Consumers' Desire for the Brand," *Journal of Consumer Research* 41, no. 3 (2014): 590–609, https://doi.org/10.1086/676980.

18. Julia D. Hur, Minjung Koo, and Wilhelm Hofmann, "When Temptations Come Alive: How Anthropomorphism Undermines Self-Control," *Journal of Consumer Research* 42, no. 2 (2015): 340–58, https://doi.org/10.1093/jcr/ucv017.

19. Ellen Berscheid, Mark Snyder, and Allen M. Omoto, "The Relationship Closeness Inventory: Assessing the Closeness of Interpersonal Relationships," *Journal of Personality and Social Psychology* 57, no. 5 (1989): 792–807, https://doi.org/10.1037/0022-3514.57.5.792.

20. Elaine Hatfield and Richard L. Rapson, *Love, Sex, and Intimacy: Their Psychology, Biology, and History* (New York: HarperCollins, 1993), 9.

21. Ahuvia, "I Love It!"

22. Aruna Ranganathan, "The Artisan and His Audience: Identification with Work and Price Setting in a Handicraft Cluster in Southern India," *Administrative Science Quarterly* 63, no. 3 (2018): 637–67, https://doi.org/10.1177/0001839217725782.

23. Irene Consiglio et al., "Brand (In)fidelity: When Flirting with the Competition Strengthens Brand Relationships" (presentation, Brands and Brand Relationships conference, Boston, May 20, 2014).

24. Oscar Ybarra, David Seungjae Lee, and Richard Gonzalez, "Supportive Social Relationships Attenuate the Appeal of Choice," *Psychological Science* 23, no. 10 (2012): 1186–92, https://doi.org/10.1177/0956797612440458.

25. Kristina M. Durante and Ashley Rae Arsena, "Playing the Field: The Effect of Fertility on Women's Desire for Variety," *Journal of Consumer Research* 41, no. 6 (2015): 1372–91, https://doi.org/10.1086/679652.

26. Aaker, Fournier, and Brasel, "When Good Brands Do Bad."

27. Claudio Alvarez and Susan Fournier, "Brand Flings: When Great Brand Relationships Are Not Made to Last," in *Consumer-Brand Relationships: Theory and Practice*, ed. Susan Fournier, Michael Breazeale, and Marc Fetscherin (Abingdon, UK: Routledge, 2013). Also see Jill Avery, Susan Fournier, and John Wittenbraker, "Unlock the Mysteries of Your Customer Relationships," *Harvard Business Review*, July–August 2014.

28. Aaker, Fournier, and Brasel, "When Good Brands Do Bad."

29. Vanitha Swaminathan, Karen M. Stilley, and Rohini Ahluwalia, "When Brand Personality Matters: The Moderating Role of Attachment Styles," *Journal of Consumer Research* 35, no. 6 (2009): 985–1002, https://doi.org/10.1086/593948.

30. Daniel Kahneman, *Thinking, Fast and Slow* (New York: Farrar, Straus and Giroux, 2011).

CHAPTER 4: People Connectors

1. Linda Wertheimer, "The Soul of the World's Most Expensive Violin," *Morning Edition*, radio broadcast, March 7, 2014, https://www.npr.org/sections/deceptivecadence/2014/03/07/286262067/the-soul-of-the-worlds-most-expensive-violin.
2. Russell W. Belk, "Possessions and the Extended Self," *Journal of Consumer Research* 15, no. 2 (1988): 139–68, https://doi.org/10.1086/209154.
3. I heard the story on the National Public Radio show *All Things Considered*. Regrettably, it aired more than twenty years ago, and despite searching, I've been unable to find the original story, so alas, I can't properly credit the reporter.
4. Aaron Chaim Ahuvia, "Nothing Matters More to People Than People: Brand Meaning and Social Relationships," *Review of Marketing Research* 12 (May 2015): 121–49, https://doi.org/10.1108/S1548-643520150000012005; Aaron Ahuvia, "Beyond 'Beyond the Extended Self': Russ Belk on Identity," in *Legends in Consumer Behavior: Russell W. Belk*, ed. Jagdish N. Sheth, vol. 4, *Consumer Sense of Self and Identity*, ed. John W. Schouten (Thousand Oaks, CA: Sage Publishing, 2014).
5. A few of these studies include Mihaly Csikszentmihalyi and Eugene Rochberg-Halton, *The Meaning of Things: Domestic Symbols and the Self* (Cambridge, MA: Cambridge University Press, 1981); Elizabeth C. Hirschman and Priscilla A. LaBarbera, "Dimensions of Possession Importance," *Psychology & Marketing* 7, no. 3 (1990): 215–33, https://doi.org/10.1002/mar.4220070306; Raj Mehta and Russell W. Belk, "Artifacts, Identity, and Transition: Favorite Possessions of Indians and Indian Immigrants to the United States," *Journal of Consumer Research* 17, no. 4 (1991): 398–411, https://doi.org/10.1086/208566; Susan E. Schultz, Robert E. Kleine, and Jerome B. Kernan, "These Are a Few of My Favorite Things: Toward an Explication of Attachment as a Consumer Behavior Construct," in *Advances in Consumer Research* 16, no. 1 (1989): 359–66; and Melanie Wallendorf and Eric Arnould, "'My Favorite Things': A Cross-Cultural Inquiry into Object

Attachment, Possessiveness, and Social Linkage," *Journal of Consumer Research* 14, no. 4 (1988): 531–47, https://doi.org/10.1086/209134.

6. Csikszentmihalyi and Rochberg-Halton, *The Meaning of Things*.

7. Aaron Ahuvia, "I Love It! Towards a Unifying Theory of Love Across Diverse Love Objects," abridged (PhD diss., Northwestern University, 1993), https://deepblue.lib.umich.edu/handle/2027.42/35351.

8. Ahuvia, "I Love It!"

9. Vanitha Swaminathan, Karen M. Stilley, and Rohini Ahluwalia, "When Brand Personality Matters: The Moderating Role of Attachment Styles," *Journal of Consumer Research* 35, no. 6 (2009): 985–1002, https://doi.org/10.1086/593948.

10. Aaron C. Ahuvia et al., "Pride of Ownership: An Identity-Based Model," *Journal of the Association for Consumer Research* 3, no. 2 (April 2018): 1–13, https://doi.org/10.1086/697076.

11. Mansur Khamitov, Miranda Goode, and Matthew Thomson, "Investigating Brand Cheating in Consumer-Brand Relationships: Triadic and Dyadic Approaches," *Advances in Consumer Research* 42 (2014): 541; Miranda Goode, Mansur Khamitov, and Matthew Thomson, "Dyads, Triads and Consumer Treachery: When Interpersonal Connections Guard Against Brand Cheating," in *Strong Brands, Strong Relationships*, ed. Susan Fournier, Michael J. Breazeale, and Jill Avery (Abingdon, UK: Routledge, 2015), 216–32.

12. Goode, Khamitov, and Thomson, "Dyads, Triads and Consumer Treachery"; Khamitov, Goode, and Thomson, "Investigating Brand Cheating in Consumer-Brand Relationships."

13. Rik Pieters, "Bidirectional Dynamics of Materialism and Loneliness: Not Just a Vicious Cycle," *Journal of Consumer Research* 40, no. 4 (2013): 615–31, https://doi.org/10.1086/671564.

14. Xijing Wang and Eva G. Krumhuber, "The Love of Money Results in Objectification," *British Journal of Social Psychology* 56, no. 2 (September 2016): 354–72, https://doi.org/10.1111/bjso.12158.

15. Monica Perez interviewed by Sam Sanders, "West Coast on Fire, Plus Comedian Sam Jay," September 11, 2020, *It's Been a Minute with Sam Sanders*, podcast, https://www.npr.org/2020/09/11/911947429/west-coast-on-fire-plus-comedian-sam-jay.

16. Aaron C. Ahuvia, "Beyond the Extended Self: Loved Objects and Consumers' Identity Narratives," *Journal of Consumer Research* 32, no. 1 (2005): 171–84, https://doi.org/10.1086/429607.

17. Marsha L. Richins, "Measuring Emotions in the Consumption Experience," *Journal of Consumer Research* 24, no. 2 (1997): 127–46, https://doi.org/10.1086/209499.

18. Cindy Chan, Jonah Berger, and Leaf Van Boven, "Identifiable but Not Identical: Combining Social Identity and Uniqueness Motives in Choice," *Journal of Consumer Research* 39, no. 3 (2012): 561–73, https://doi.org/10.1086/664804.

19. For an excellent discussion of the ways in which entertainment products help us create social relationships, see Cristel Antonia Russell and Hope Jensen Schau, "When Narrative Brands End: The Impact of Narrative Closure and Consumption Sociality on Loss Accommodation," *Journal of Consumer Research* 40, no. 6 (2014): 1039–62, https://doi.org/10.1086/673959.

20. Some relevant studies on brand communities include Richard P. Bagozzi et al., "Customer-Organization Relationships: Development and Test of a Theory of Extended Identities," *Journal of Applied Psychology* 97, no. 1 (2012): 63–76, https://doi.org/10.1037/a0024533; Richard P. Bagozzi and Utpal M. Dholakia, "Antecedents and Purchase Consequences of Customer Participation in Small Group Brand Communities," *International Journal of Research in Marketing* 23, no. 1 (2006): 45–61, https://doi.org/10.1016/j.ijresmar.2006.01.005; Lars Bergkvist and Tino Bech-Larsen, "Two Studies of Consequences and Actionable Antecedents of Brand Love," *Journal of Brand Management* 17, no. 7 (2010): 504–18, http://doi.org/10.1057/bm.2010.6; Bernard Cova, "Community and Consumption: Towards a Definition of the 'Linking Value' of Product or Services," *European Journal of Marketing* 31, no. 3/4 (1997), 297–316, https://doi.org/10.1108/03090569710162380; Hope Jensen Schau, Albert M. Muñiz, and Eric J. Arnould, "How Brand Community Practices Create Value," *Journal of Marketing* 73, no. 5 (2009): 30–51, https://doi.org/10.1509/jmkg.73.5.30; and Cleopatra Veloutsou and Luiz Moutinho, "Brand Relationships Through Brand Reputation and Brand Tribalism," *Journal of Business Research* 62, no. 3 (2009): 314–22, https://doi.org/10.1016/j.jbusres.2008.05.010.

21. Angela Watercutter, "Brony Census Tracks 'State of the Herd,'" *Wired*, January 10, 2012, https://www.wired.com/2012/01/brony-census/.

22. Logan Hamley et al., "Ingroup Love or Outgroup Hate (or Both)? Mapping Distinct Bias Profiles in the Population," *Personality and*

Social Psychology Bulletin 46, no. 2 (2020): 171–88, https://doi .org/10.1177/0146167219845919.

23. Maja Golf Papez and Michael Beverland, "Exploring the Negative Aspects of Consumer Brand Relationships Through the Use of Relational Models Theory" (presentation, Brands and Brand Relationships conference, Toronto, May 20, 2016).

24. Aaron C. Ahuvia, Rajeev Batra, and Richard P. Bagozzi, "Love, Desire and Identity: A Conditional Integration Theory of the Love of Things," in *The Handbook of Brand Relationships*, ed. Deborah J. MacInnis, C. Whan Park, and Joseph R. Priester (New York: M. E. Sharpe, 2009).

25. Adam C. Landon et al., "Psychological Needs Satisfaction and Attachment to Natural Landscapes," *Environment and Behavior* 53, no. 6 (2020): 661–83, https://doi.org/10.1177/0013916520916255.

26. Data collected for an ongoing project.

CHAPTER 5: **You Are What You Love**

1. Aaron Ahuvia, "I Love It! Towards a Unifying Theory of Love Across Diverse Love Objects," abridged (PhD diss., Northwestern University, 1993), https://deepblue.lib.umich.edu/handle/2027.42/35351.

2. Data collected for Aaron C. Ahuvia, "Beyond the Extended Self: Loved Objects and Consumers' Identity Narratives," *Journal of Consumer Research* 32, no. 1 (2005): 171–84, https://doi.org/10.1086/429607.

3. Arthur Aron et al., "Close Relationships as Including Other in the Self," *Journal of Personality and Social Psychology* 60, no. 2 (1991): 241–53, https://doi.org/10.1037/0022-3514.60.2.241.

4. Arthur Aron and Barbara Fraley, "Relationship Closeness as Including Other in the Self: Cognitive Underpinnings and Measures," *Social Cognition* 17, no. 2 (1999): 140–60, https://doi.org/10.1521/soco .1999.17.2.140.

5. C. Whan Park, Andreas B. Eisingerich, and Jason Whan Park, "From Brand Aversion or Indifference to Brand Attachment: Authors' Response to Commentaries to Park, Eisingerich, and Park's Brand Attachment-Aversion Model," *Journal of Consumer Psychology* 23, no. 2 (2013): 269–74, https://doi.org/10.1016/j.jcps.2013.01.006.

6. Sara H. Konrath and Michael Ross, "Our Glories, Our Shames: Expanding the Self in Temporal Self Appraisal Theory" (conference poster

presented at the 111th annual meeting of the American Psychological Society, Atlanta, August 2003), http://hdl.handle.net/1805/10039.

7. Shinya Watanuki and Hiroyuki Akama, "Neural Substrates of Brand Love: An Activation Likelihood Estimation Meta-Analysis of Functional Neuroimaging Studies," *Frontiers in Neuroscience* 14 (2020), https://doi.org/10.3389/fnins.2020.534671.

8. Rajeev Batra, Aaron Ahuvia, and Richard P. Bagozzi, "Brand Love," *Journal of Marketing* 76, no. 2 (2012): 1–16, https://doi.org/10.1509/jm.09.0339; Richard P. Bagozzi, Rajeev Batra, and Aaron Ahuvia, "Brand Love: Development and Validation of a Practical Scale," *Marketing Letters* 28 (2016): 1–14, https://doi.org/10.1007/s11002-016-9406-1.

9. Judy A. Shea and Gerald R. Adams, "Correlates of Romantic Attachment: A Path Analysis Study," *Journal of Youth and Adolescence* 13, no. 1 (1984): 27–44, https://doi.org/10.1007/BF02088651.

10. Data collected for Ahuvia, "Beyond the Extended Self."

11. Pamela Paul, "Jeffrey Toobin on Writing About Trump," September 4, 2020, *New York Times Book Review Podcast*, 11:33, https://www.nytimes.com/2020/09/04/books/review/podcast-jeffrey-toobin-true-crimes-misdemeanors-trump-dayna-tortorici-elena-ferrante.html.

12. Ahuvia, "Beyond the Extended Self."

13. Data collected for Aaron C. Ahuvia et al., "Pride of Ownership: An Identity-Based Model," *Journal of the Association for Consumer Research* 3, no. 2 (April 2018): 1–13, https://doi.org/10.1086/697076.

14. Ahuvia, "I Love It!"

15. Elizabeth Mehren, "Oh, Jackie! What Next? They've Got Big Plans for Those Pricey Buys," *Los Angeles Times*, June 13, 1996, https://www.latimes.com/archives/la-xpm-1996-06-13-ls-14295-story.html.

16. Carol J. Nemeroff and Paul Rozin, "The Contagion Concept in Adult Thinking in the United States: Transmission of Germs and of Interpersonal Influence," *Ethos* 22, no. 2 (2009): 158–86, https://doi.org/10.1525/eth.1994.22.2.02a00020.

17. Jennifer J. Argo, Darren W. Dahl, and Andrea C. Morales, "Positive Consumer Contagion: Responses to Attractive Others in a Retail Context," *Journal of Marketing Research* 45, no. 6 (2008): 690–701, https://doi.org/10.1509/jmkr.45.6.690.

18. Chris Speed, "From RememberMe to Shelflife," *Fields*, February 27, 2012, http://www.chrisspeed.net/?p=773.

19. Erich Fromm, *The Art of Loving: An Enquiry into the Nature of Love* (New York: Harper & Brothers, 1956), 17.

20. Michael I. Norton, Daniel Mochon, and Dan Ariely, "The IKEA Effect: When Labor Leads to Love," *Journal of Consumer Psychology* 22, no. 3 (2012): 453–60, https://doi.org/10.1016/j.jcps.2011.08.002.

21. Peter H. Bloch, "Involvement Beyond the Purchase Process: Conceptual Issues and Empirical Investigation," *Advances in Consumer Research* 9, no. 1 (1982): 413–17.

22. Russell W. Belk, "Possessions and Extended Sense of Self," in *Marketing and Semiotics: New Directions in the Study of Signs for Sale*, ed. Jean Umikeer-Sebeok (Berlin: Mouton de Gruyter, 1987), 151–64.

23. Norton, Mochon, and Ariely, "The IKEA Effect."

24. Data collected for Aaron C. Ahuvia, Rajeev Batra, and Richard P. Bagozzi, "Love, Desire and Identity: A Conditional Integration Theory of the Love of Things," in *The Handbook of Brand Relationships*, ed. Deborah J. MacInnis, C. Whan Park, and Joseph R. Priester (New York: M. E. Sharpe, 2009).

25. Arthur Aron, Meg Paris, and Elaine N. Aron, "Falling in Love: Prospective Studies of Self-Concept Change," *Journal of Personality and Social Psychology* 69, no. 6 (1995): 1102–12, https://doi.org/10.1037/0022-3514.69.6.1102.

26. William James, *The Principles of Psychology* (New York: Henry Holt, 1890), 1:291.

27. Lea Dunn and JoAndrea Hoegg, "The Impact of Fear on Emotional Brand Attachment," *Journal of Consumer Research* 41, no. 1 (2014): 152–68, https://doi.org/10.1086/675377.

28. Sarah Broadbent, "Brand Love in Sport: Antecedents and Consequences" (PhD diss., School of Management and Marketing, Deakin University, 2012), https://dro.deakin.edu.au/view/DU:30062512.

CHAPTER 6: Finding Ourselves in the Things We Love

1. Mason Haire, "Projective Techniques in Marketing Research," *Journal of Marketing* 14, no. 5 (April 1950): 649–56, https://doi.org/10.2307/1246942.

2. Adam Smith, *The Wealth of Nations* (London: W. Strahan and T. Cadell, 1776), vol. 2, bk. 5, ch. 2, https://www.marxists.org/reference archive/smith-adam/works/wealth-of-nations/book05/ch02b-4.htm.

3. Tori DeAngelis, "A Theory of Classism: Class Differences," *Monitor on Psychology* 46, no. 2 (2015): 62.

4. Ronald Inglehart, *Culture Shift in Advanced Industrial Society* (Princeton, NJ: Princeton University Press, 1990); Aaron C. Ahuvia and Nancy Y. Wong, "Materialism: Origins and Implications for Personal Well-Being," *European Advances in Consumer Research* 2 (1995): 172–78; Aaron C. Ahuvia and Nancy Y. Wong, "Personality and Values-Based Materialism: Their Relationship and Origins," *Journal of Consumer Psychology* 12, no. 4 (2002): 389–402, https://doi.org/10.1016/S1057-7408(16)30089-4; Nancy Wong and Aaron Chaim Ahuvia, "Personal Taste and Family Face: Luxury Consumption in Confucian and Western Societies," *Psychology & Marketing* 15, no. 5 (1998): 423–41, https://doi.org/10.1002/(SICI)1520-6793(199808)15:5<423::AID-MAR2>3.0.CO;2-9.

5. Aaron C. Ahuvia, "Individualism/Collectivism and Cultures of Happiness: A Theoretical Conjecture on the Relationship Between Consumption, Culture and Subjective Well-Being at the National Level," *Journal of Happiness Studies* 3, no. 1 (2002): 23–36, http://dx.doi.org/10.1023/A:1015682121103.

6. Data collected for Wong and Ahuvia, "Personal Taste and Family Face."

7. Hazel R. Markus and Shinobu Kitayama, "Culture and the Self: Implications for Cognition, Emotion, and Motivation," *Psychological Review* 98, no. 2 (1991): 224–53, https://doi.org/10.1037/0033-295X.98.2.224.

8. Data collected for Wong and Ahuvia, "Personal Taste and Family Face."

9. Ahuvia, "Individualism/Collectivism and Cultures of Happiness."

10. Shankar Vedantam, "You 2.0: Loss and Renewal," *Hidden Brain*, radio broadcast, August 17, 2020, https://www.npr.org/2020/08/15/902891952/you-2-0-loss-and-renewal.

11. Data collected for Rajeev Batra, Aaron Ahuvia, and Richard P. Bagozzi, "Brand Love," *Journal of Marketing* 76, no. 2 (2012): 1–16, https://doi.org/10.1509/jm.09.0339.

12. Data collected for Batra, Ahuvia, and Bagozzi, "Brand Love."

13. Why don't the two figures — 80 percent and 10 percent — add up to 100 percent? Because some people mention neither. The data comes from Aaron Ahuvia, "I Love It! Towards a Unifying Theory of Love Across Diverse Love Objects," abridged (PhD diss., Northwestern University, 1993), https://deepblue.lib.umich.edu/handle/2027.42/35351.

14. Data collected for Aaron C. Ahuvia, "Beyond the Extended Self: Loved Objects and Consumers' Identity Narratives," *Journal of Consumer Research* 32, no. 1 (2005): 171–84, https://doi.org/10.1086/429607.

15. For another example of this, see Alina Selyukh, "She Works Two Jobs. Her Grocery Budget Is $25. This Is Life Near Minimum Wage," *All Things Considered,* radio broadcast, March 25, 2021, https://www.npr.org/2021/03/26/979983739/walk-one-day-in-our-shoes-life-near-minimum-wage.

16. Nathaniel Branden, "A Vision of Romantic Love," in *The Psychology of Love,* ed. Robert Sternberg and Michael L. Barnes (New Haven, CT: Yale University Press, 1988), 224.

17. Data collected for Batra, Ahuvia, and Bagozzi, "Brand Love."

18. Ahuvia, "I Love It!"

19. "Cartoonist Cathy Guisewite on Her Best Gift Ever," *Marketplace,* radio broadcast, December 15, 2014, https://www.marketplace.org/2014/12/15/cartoonist-cathy-guisewite-her-best-gift-ever/.

20. Ahuvia, "Beyond the Extended Self."

21. Data collected for Ahuvia, "Beyond the Extended Self."

22. Ahuvia, "Beyond the Extended Self."

23. Wong and Ahuvia, "Personal Taste and Family Face."

24. Data collected for Wong and Ahuvia, "Personal Taste and Family Face."

25. Lynn Hirschberg, "Next. Next. What's Next?," *New York Times Magazine,* April 7, 1996.

CHAPTER 7: **Enjoyment and Flow**

1. Julie A. Mennella, Coren P. Jagnow, and Gary K. Beauchamp, "Prenatal and Postnatal Flavor Learning by Human Infants," *Pediatrics* 107, no. 6 (2001): e88, https://doi.org/10.1542/peds.107.6.e88.

2. This observation was made by David Rosen while appearing as a guest on the podcast *Inquiring Minds* in an episode titled "Generating the Element of Harmonic Surprise with David Rosen" (July 12, 2021).

3. Robert M. Sapolsky, "Open Season," *The New Yorker,* March 22, 1998.

4. Unless otherwise cited, the information in this and subsequent paragraphs comes from Hilary Coon and Gregory Carey, "Genetic and Environmental Determinants of Musical Ability in Twins," *Behavior Genetics* 19 (March 1989): 183–93, https://doi.org/10.100/BF01065903.

5. Jakob Pietschnig and Martin Voracek, "One Century of Global IQ Gains: A Formal Meta-Analysis of the Flynn Effect (1909–2013)," *Perspectives on Psychological Science* 10, no. 3 (May 2015): 282–306, https://doi.org/10.1177/1745691615577701.

6. Madison Troyer, "Baby Names Gaining Popularity in the 21st Century," *Stacker*, April 24, 2021.

7. Madison Troyer, "Baby Names Losing Popularity in the 21st Century," *Stacker*, April 8, 2021.

8. Maria A. Rodas and Carlos J. Torelli, "The Self-Expanding Process of Falling in Love with a Brand" (presentation, Brands and Brand Relationships conference, Toronto, May 20, 2016).

9. Felix Richter, "Gaming: The Most Lucrative Entertainment Industry by Far," Statista, September 22, 2020, https://www.statista.com/chart/22392/global-revenue-of-selected-entertainment-industry-sectors/.

10. Luke Appleby, "Gabe Newell Says Brain-Computer Interface Tech Will Allow Video Games Far Beyond What Human 'Meat Peripherals' Can Comprehend," 1 NEWS, January 24, 2021.

11. James J. Kellaris and Ronald C. Rice, "The Influence of Tempo, Loudness, and Gender of Listener on Responses to Music," *Psychology & Marketing* 10, no. 1 (1993): 15–29, https://doi.org/10.1002/mar.4220100103.

12. Kendra Cherry, "When and Why Does Habituation Occur?," *Verywell Mind*, December 2, 2020.

13. "U.S. Adult Consumption of Added Sugars Increased by More Than 30% over Three Decades," *ScienceDaily*, November 4, 2014, www.sciencedaily.com/releases/2014/11/141104141731.htm.

14. Chiadi E. Ndumele, "Obesity, Sugar and Heart Health," Johns Hopkins Medicine, https://www.hopkinsmedicine.org/health/wellness-and-prevention/obesity-sugar-and-heart-health.

15. Office of Public Affairs, University of Utah Health, "Sweet Nothings: Added Sugar Is a Top Driver of Diabetes," February 10, 2015, https://healthcare.utah.edu/healthfeed/postings/2015/02/021015_cvarticle-sugar-diabetes.php#:~:text=They%20found%20that%20added%20sugar,to%20inflammation%20and%20insulin%20resistance.

16. Data collected for Aaron C. Ahuvia, "Beyond the Extended Self: Loved Objects and Consumers' Identity Narratives," *Journal of Consumer Research* 32, no. 1 (2005): 171–84, https://doi.org/10.1086/429607.

17. Data collected for Aaron C. Ahuvia, Rajeev Batra, and Richard P. Bagozzi, "Love, Desire and Identity: A Conditional Integration Theory of the Love of Things," in *The Handbook of Brand Relationships*, ed. Deborah J. MacInnis, C. Whan Park, and Joseph R. Priester (New York: M. E. Sharpe, 2009).

18. "How the Nose Knows," June 18, 2021, *The Pulse*, produced by Maiken Scott, podcast, https://whyy.org/episodes/how-the-nose-knows/.

19. "How the Nose Knows."

20. Stephen E. Palmer and Karen B. Schloss, "An Ecological Valence Theory of Human Color Preference," *Proceedings of the National Academy of Sciences of the United States of America* 107, no. 19 (2010): 8877–82, https://doi.org/10.1073/pnas.0906172107.

CHAPTER 8: **What the Things We Love Say About Us**

1. Omri Gillath et al., "Shoes as a Source of First Impressions," *Journal of Research in Personality* 46, no. 4 (2012): 423–30, https://doi.org/10.1016/j.jrp.2012.04.003.

2. Data collected for Nancy Wong and Aaron Chaim Ahuvia, "Personal Taste and Family Face: Luxury Consumption in Confucian and Western Societies," *Psychology & Marketing* 15, no. 5 (1998): 423–41, https://doi.org/10.1002/(SICI)1520-6793(199808)15:5<423::AID-MAR2>3.0.CO;2-9.

3. Information on the Mosaic lifestyle groups is available on the website of the library of the University of Texas at El Paso: https://libguides.utep.edu/comm_yang/Demographics_Now_Mosaic_Clusters.

4. My thinking on this topic was particularly influenced by Douglas B. Holt, "Does Cultural Capital Structure American Consumption?," *Journal of Consumer Research* 25, no. 1 (1998): 1–25, https://doi.org/10.1086/209523.

5. The theory that underlies most of my observations in this chapter comes from Pierre Bourdieu. His classic work on this topic is *Distinction: A Social Critique of the Judgement of Taste* (Cambridge, MA: Harvard University Press, 1987).

6. Aaron C. Ahuvia et al., "What Is the Harm in Fake Luxury Brands? Moving Beyond the Conventional Wisdom," in *Luxury Marketing: A Challenge for Theory and Practice*, ed. Klaus-Peter Wiedmann and

Nadine Hennigs (Wiesbaden: Gabler Verlag, 2012), 279–93, https://doi.org/10.1007/978-3-8349-4399-6_16.

7. Tori DeAngelis, "A Theory of Classism: Class Differences," *Monitor on Psychology* 46, no. 2 (2015): 62. Also see Antony S. R. Manstead, "The Psychology of Social Class: How Socioeconomic Status Impacts Thought, Feelings, and Behaviour," *British Journal of Social Psychology* 57, no. 2 (2018): 267–91, https://doi.org/10.1111/bjso.12251.

8. Hazel R. Markus and Shinobu Kitayama, "Culture and the Self: Implications for Cognition, Emotion, and Motivation," *Psychological Review* 98, no. 2 (1991): 224–53, https://doi.org/10.1037/0033-295X.98.2.224.

9. Nicole M. Stephens, Hazel Rose Markus, and L. Taylor Phillips, "Social Class Culture Cycles: How Three Gateway Contexts Shape Selves and Fuel Inequality," *Annual Review of Psychology* 65 (2014): 611–34, https://doi.org/10.1146/annurev-psych-010213-115143.

10. Cathy Horyn, "Yves Saint Laurent Assembles a 'New Tribe,'" *New York Times*, October 5, 2010, https://runway.blogs.nytimes.com/2010/10/05/yves-saint-laurent-assembles-a-new-tribe/?searchResultPosition=1.

11. If you're interested in cultural capital or the nature of good taste versus bad taste, I highly recommend Carl Wilson's *Let's Talk About Love: Why Other People Have Such Bad Taste* (New York: Bloomsbury, 2014).

12. Robert M. Lupton, Steven M. Smallpage, and Adam M. Enders, "Values and Political Predispositions in the Age of Polarization: Examining the Relationship Between Partisanship and Ideology in the United States, 1988–2012," *British Journal of Political Science* 50, no. 1 (2020): 241–60, https://doi.org/10.1017/S0007123417000370.

13. Judith Martin, "Shock Your Dinner Guests: Give 'Em the Asparagus Rule," *Chicago Tribune*, April 18, 2007.

14. Richard A. Peterson and Roger M. Kern, "Changing Highbrow Taste: From Snob to Omnivore," *American Sociological Review* 61, no. 5 (1996): 900–907, https://doi.org/10.2307/2096460.

15. For a still timely and highly entertaining account of this, see David Brooks, *Bobos in Paradise: The New Upper Class and How They Got There* (New York: Simon and Schuster, 2000).

16. Holt, "Does Cultural Capital Structure American Consumption?"

17. Jane Coaston, "Is Fox News Really All That Powerful?," June 30, 2021, *The Argument*, podcast, 9:51, https://www.nytimes.com/2021/06/30/opinion/power-politics-culture-war.html.

18. Paul Henry Ray and Sherry Ruth Anderson, *The Cultural Creatives: How 50 Million People Are Changing the World* (New York: Three Rivers Press, 2000). This group has also been called the "creative class." See Richard Florida, *The Rise of the Creative Class* (New York: Basic Books, 2002); and Brooks, *Bobos in Paradise*.

19. Data collected for Wong and Ahuvia, "Personal Taste and Family Face."

20. Mary Douglas and Baron C. Isherwood, *The World of Goods: Towards an Anthropology of Consumption* (New York: Basic Books, 1979), 85.

21. For an excellent paper that influenced my thinking on this, see Young Jee Han, Joseph C. Nunes, and Xavier Drèze, "Signaling Status with Luxury Goods: The Role of Brand Prominence," *Journal of Marketing* 74, no. 4 (2010): 15–30, https://doi.org/10.1509/jmkg.74.4.015.

22. Ahuvia et al., "What Is the Harm in Fake Luxury Brands?"

23. Data collected for Wong and Ahuvia, "Personal Taste and Family Face."

24. From unpublished data collected for a commercial research project.

25. Jack Houston and Irene Anna Kim, "Why Hermès Birkin Bags Are So Expensive, According to a Handbag Expert," *Business Insider*, June 30, 2021, https://www.businessinsider.com/hermes-birkin-bag-realreal-handbag-expert-so-expensive-2019-6?amp.

26. Data collected for Wong and Ahuvia, "Personal Taste and Family Face."

CHAPTER 9: Because Evolution

1. David Gal, "A Mouth-Watering Prospect: Salivation to Material Reward," *Journal of Consumer Research* 38, no 6 (2012): 1022–29, https://doi.org/10.1086/661766.

2. These first three stages are not to be confused with Helen Fisher's three brain systems, which underlie mate choice. There is no fundamental conflict between my approach and Fisher's. We simply draw the boundary lines between stages in different places, because I want to highlight some aspects of love that are particularly relevant to the love of things. Specifically, I (and psychologists generally) break Fisher's "attraction system" into two subsystems: one for identifying mates with attractive genes and another for identifying mates likely to be good parents. Fisher and her colleagues discussed some of the reasons why psychologists focus attention on these two subsystems, as well as some evidence from biologists that make the same distinction,

in Helen Fisher et al., "The Neural Mechanisms of Mate Choice: A Hypothesis," *Neuro Endocrinology Letters* 23 Suppl. 4 (2002): 92–97.

3. Beverley Fehr, "How Do I Love Thee? Let Me Consult My Prototype," in *Individuals in Relationships*, ed. Steve Duck (Newbury Park, CA: Sage Publications, 1993), 87–120, http://dx.doi.org/10.4135/9781483326283.n4.

4. Lawrence S. Sugiyama, "Physical Attractiveness in Adaptationist Perspective," in *The Handbook of Evolutionary Psychology*, ed. David M. Buss (Hoboken, NJ: John Wiley & Sons, 2005), 292–343, https://doi.org/10.1002/9780470939376.ch10.

5. Data collected for Aaron C. Ahuvia, Rajeev Batra, and Richard P. Bagozzi, "Love, Desire and Identity: A Conditional Integration Theory of the Love of Things," in *The Handbook of Brand Relationships*, ed. Deborah J. MacInnis, C. Whan Park, and Joseph R. Priester (New York: M. E. Sharpe, 2009).

6. Claudia Townsend and Sanjay Sood, "Self-Affirmation Through the Choice of Highly Aesthetic Products," *Journal of Consumer Research* 39, no. 2 (2012): 415–28, https://doi.org/10.1086/663775.

7. Jean-Jacques Rousseau, *A Discourse on Inequality*, trans. Maurice Cranston (New York: Viking, 1984), 167.

8. Helen Fisher, Arthur Aron, and Lucy L. Brown, "Romantic Love: An fMRI Study of a Neural Mechanism for Mate Choice," *Journal of Comparative Neurology* 493, no. 1 (2005): 58–62, https://doi.org/10.1002/cne.20772.

9. Fisher, Aron, and Brown, "Romantic Love."

10. Jon Hamilton, "From Primitive Parts, a Highly Evolved Human Brain," *Morning Edition*, radio broadcast, August 9, 2010, http://www.npr.org/templates/story/story.php?storyId=129027124.

11. Claudio Alvarez and Susan Fournier, "Brand Flings: When Great Brand Relationships Are Not Made to Last," in *Consumer-Brand Relationships: Theory and Practice*, ed. Susan Fournier, Michael Breazeale, and Marc Fetscherin (Abingdon, UK: Routledge, 2013).

12. Helen Fisher et al., "Defining the Brain Systems of Lust, Romantic Attraction, and Attachment," *Archives of Sexual Behavior* 31, no. 5 (2002), 413–19, https://doi.org/ 10.1023/a:1019888024255.

13. Sara M. Freeman and Larry J. Young, "Oxytocin, Vasopressin, and the Evolution of Mating Systems in Mammals," in *Oxytocin, Vasopressin, and Related Peptides in the Regulation of Behavior*, ed. Elena

Choleris, Donald W. Pfaff, and Martin Kavaliers (Cambridge, UK: Cambridge University Press, 2013), 128–47.

14. Fisher et al., "Defining the Brain Systems of Lust, Romantic Attraction, and Attachment."

15. Aaron Ahuvia, "I Love It! Towards a Unifying Theory of Love Across Diverse Love Objects," abridged (PhD diss., Northwestern University, 1993), https://deepblue.lib.umich.edu/handle/2027.42/35351.

16. Bernard I. Murstein, "Mate Selection in the 1970s," *Journal of Marriage and the Family* 42, no. 4 (1980), 777–92, https://doi.org/10.2307/351824.

17. Aaron C. Ahuvia and Mara B. Adelman, "Market Metaphors for Meeting Mates," in *Research in Consumer Behavior* vol. 6, ed. Janeen A. Costa and Russell W. Belk (Greenwich, CT: JAI Press, 1993), 55–83.

18. Larry Young and Brian Alexander, "Be My Territory," in *The Chemistry Between Us: Love, Sex, and the Science of Attraction* (New York: Current, 2012), 154–84. Also see Freeman and Young, "Oxytocin, Vasopressin, and the Evolution of Mating Systems in Mammals"; and Hasse Walum and Larry H. Young, "The Neural Mechanisms and Circuitry of the Pair Bond," *Nature Reviews Neuroscience* 19, no. 11 (2018): 643–54, https://doi.org/10.1038/s41583-018-0072-6.

19. R. I. M. Dunbar and Susanne Shultz, "Evolution in the Social Brain," *Science* 317, no. 5843 (2007): 1344–47, https://doi.org/10.1126/science.1145463.

20. Hamilton, "From Primitive Parts, a Highly Evolved Human Brain."

21. Dunbar and Shultz, "Evolution in the Social Brain."

22. Garth J. O. Fletcher et al., "Pair-Bonding, Romantic Love, and Evolution: The Curious Case of *Homo sapiens*," *Perspectives on Psychological Science* 10, no. 1 (2015): 20–36, https://doi.org/10.1177/1745691614561683.

23. Arthur Aron, Elaine N. Aron, and Danny Smollan, "Inclusion of Other in the Self Scale and the Structure of Interpersonal Closeness," *Journal of Personality and Social Psychology* 63, no. 4 (1992): 596–612, https://doi.org/10.1037/0022-3514.63.4.596.

24. Stephen J. Dollinger and Stephanie M. Clancy, "Identity, Self, and Personality: II. Glimpses Through the Autophotographic Eye," *Journal of Personality and Social Psychology* 64, no. 6 (1993): 1064–71, https://doi.org/10.1037/0022-3514.64.6.1064.

25. The fusiform face area is part of a larger area of the brain called the fusiform gyrus. The scientific literature sometimes uses the term "fusiform gyrus" synonymously with the term "fusiform face area."

26. Josef Parvizi et al., "Electrical Stimulation of Human Fusiform Face-Selective Regions Distorts Face Perception," *Journal of Neuroscience* 32, no. 43 (October 2012): 14915–20, https://doi.org/10.1523 /JNEUROSCI.2609-12.2012; Elizabeth Norton, "Facial Recognition: Fusiform Gyrus Brain Region 'Solely Devoted' to Faces, Study Suggests," *HuffPost*, October 24, 2012, https://www.huffpost.com/entry /facial-recognition-brain-fusiform-gyrus_n_2010192.

27. Nicolas Kervyn, Susan T. Fiske, and Chris Malone, "Brands as Intentional Agents Framework: How Perceived Intentions and Ability Can Map Brand Perception," *Journal of Consumer Psychology* 22, no. 2 (2012): 166–76, https://doi.org/10.1016/j.jcps.2011.09.006.

28. Carolyn Yoon et al., "A Functional Magnetic Resonance Imaging Study of Neural Dissociations Between Brand and Person Judgments," *Journal of Consumer Research* 33, no. 1 (2006): 31–40, https://doi.org /10.1086/504132.

29. Ken Manktelow, *Thinking and Reasoning: An Introduction to the Psychology of Reason, Judgment and Decision Making* (Hove, UK: Psychology Press, 2012).

30. Leda Cosmides and John Tooby, "Cognitive Adaptations for Social Exchange," in *The Adapted Mind: Evolutionary Psychology and the Generation of Culture*, ed. Jerome H. Barkow, Leda Cosmides, and John Tooby (New York: Oxford University Press, 1992), 163–228, https://doi.org/10.1098/rstb.2006.1991.

31. Timothy D. Wilson, *Strangers to Ourselves: Discovering the Adaptive Unconscious* (Cambridge, MA: Belknap Press of Harvard University Press, 2002).

32. Association for Psychological Science, "Harlow's Classic Studies Revealed the Importance of Maternal Contact," June 20, 2018, https:// www.psychologicalscience.org/publications/observer/obsonline /harlows-classic-studies-revealed-the-importance-of-maternal-contact .html.

33. Bronislaw Malinowski, *Argonauts of the Western Pacific: An Account of Native Enterprise and Adventure in the Archipelagoes of Melanesian New Guinea* (London: George Routledge & Sons, 1922).

34. Melanie Wallendorf and Eric Arnould, "'My Favorite Things': A Cross-Cultural Inquiry into Object Attachment, Possessiveness, and

Social Linkage," *Journal of Consumer Research* 14, no. 4 (1988): 531–47, https://doi.org/10.1086/209134.

35. Phillip Shaver et al., "Emotion Knowledge: Further Exploration of a Prototype Approach," *Journal of Personality and Social Psychology* 52, no. 6 (1987): 1061–86, https://doi.org/10.1037//0022-3514.52.6.1061.

36. Jesse Chandler and Norbert Schwarz, "Use Does Not Wear Ragged the Fabric of Friendship: Thinking of Objects as Alive Makes People Less Willing to Replace Them," *Journal of Consumer Psychology* 20, no. 2 (2010): 138–45, https://doi.org/10.1016/j.jcps.2009.12.008.

CHAPTER 10: The Future of the Things We Love

1. Alejandra Martins and Paul Rincon, "Paraplegic in Robotic Suit Kicks Off World Cup," BBC News, June 12, 2014, https://www.bbc.com/news/science-environment-27812218.

2. Ambra Sposito et al., "Extension of Perceived Arm Length Following Tool-Use: Clues to Plasticity of Body Metrics," *Neuropsychologia* 50, no. 9 (2012): 2187–94, https://doi.org/10.1016/j.neuropsychologia.2012.05.022.

3. Russell W. Belk, "Extended Self in a Digital World," *Journal of Consumer Research* 40, no. 33 (2013): 477–500, https://doi.org/10.1086/671052.

4. David Pogue, "How Far Away Is Mind-Machine Integration?," *Scientific American*, December 1, 2012, https://www.scientificamerican.com/article/how-far-away-mind-machine-integration/.

5. Theresa Machemer, "New Device Allows Man with Paralysis to Type by Imagining Handwriting," *Smithsonian*, May 14, 2021, https://www.smithsonianmag.com/smart-news/experimental-device-allows-man-paralyzed-below-neck-type-thinking-180977729/.

6. Agence France-Presse, "A Paralyzed Man's Brain Waves Converted to Speech in a World-First Breakthrough," *ScienceAlert*, July 16, 2021, https://www.sciencealert.com/scientists-have-converted-a-paralyzed-man-s-brain-waves-to-speech.

7. "Robotic 'Third Thumb' Use Can Alter Brain Representation of the Hand," University College London, May 20, 2021, https://www.ucl.ac.uk/news/2021/may/robotic-third-thumb-use-can-alter-brain-representation-hand.

8. Richard H. Passman, "Providing Attachment Objects to Facilitate Learning and Reduce Distress: Effects of Mothers and Security

Blankets," *Developmental Psychology* 13, no. 1 (1977): 25, https://doi
.org/10.1037/0012-1649.13.1.25.

9. Jodie Whelan et al., "Relational Domain Switching: Interpersonal Insecurity Predicts the Strength and Number of Marketplace Relationships," *Psychology & Marketing* 33, no. 6 (2016): 465–79, https://doi
.org/10.1002/mar.20891.

10. Marian R. Banks, Lisa M. Willoughby, and William A. Banks,
"Animal-Assisted Therapy and Loneliness in Nursing Homes: Use of
Robotic Versus Living Dogs," *Journal of the American Medical Di-
rectors Association* 9, no. 3 (2008): 173–77, https://doi.org/10.1016/j
.jamda.2007.11.007.

11. Amanda Sharkey and Noel Sharkey, "Granny and the Robots: Ethical
Issues in Robot Care for the Elderly," *Ethics and Information Technology* 14 (2012): 27–40, https://doi.org/10.1007/s10676-010-9234-6.

12. Stacey Vanek Smith, "How a Machine Learned to Spot Depression,"
Public Radio East, May 20, 2015, http://publicradioeast.org/post/how
-machine-learned-spot-depression.

13. James Vlahos, "Barbie Wants to Get to Know Your Child," *New York
Times Magazine*, September 16, 2015.

14. "Spotlight: Toyota Encourages Drivers to 'Friend' Their Cars,"
eMarketer, July 6, 2011.

15. Miguel Pais-Vieira et al., "A Brain-to-Brain Interface for Real-Time
Sharing of Sensorimotor Information," *Scientific Reports* 3, no. 1319
(2013): 1–10, https://doi.org/10.1038/srep01319.

16. Tim Urban, "Neuralink and the Brain's Magical Future," *Wait But
Why*, April 20, 2017, https://waitbutwhy.com/2017/04/neuralink.html.

17. Garth J. O. Fletcher et al., *The Science of Intimate Relationships*
(Malden, MA: Wiley Blackwell, 2013).

18. Carolyn Parkinson, Adam M. Kleinbaum, and Thalia Wheatley,
"Similar Neural Responses Predict Friendship," *Nature Communications* 9, no. 332 (2018), https://doi.org/10.1038/s41467-017-02722-7.

19. Pavel Goldstein et al., "Brain-to-Brain Coupling During Handhold-
ing Is Associated with Pain Reduction," *Proceedings of the National
Academy of Sciences of the United States of America* 115, no. 11
(2018): E2528–37, https://doi.org/10.1073/PNAS.1703643115.

20. Erich Fromm, *The Art of Loving: An Enquiry into the Nature of Love*
(New York: Harper & Brothers, 1956).

Index

Note: *Italic* page numbers refer to illustrations and charts.

Aaker, Jennifer, 82
aboriginal tribes, 255, 256
activities
    buying and owning equipment for, 131
    evolution and, 259
    finding yourself in, 143–44
    flow theory and, 173–74, 175, 176, 177, 180
    love of things shaped by relationships to, 12
    psychological visibility and, 147–49
    responsiveness and, 71
    self-concept and, 109
    sense of self and, 276
    as things, 5, 7, 8
Adelman, Mara, 4
ADHD, 182
adolescence through early adulthood, tastes in, 164–66, 167, 169
advertising
    complexity limited in, 180–81
    cultural capital and, 226
    fashion cycle and, 170
    identities and, 154–57, 155n
aesthetic formalism, 187–88
Aggarwal, Pankaj, 51–52
Ahluwalia, Rohini, 91
airline frequent-flier plans, 75
Akama, Hiroyuki, 119
Aknin, Lara, 247
alcohol, 72, 179, 239

altruism
    groups and, 243–44
    investment versus, 79–80
Alzheimer's disease, 169
Amazon, 7
anger
    anthropomorphic thinking and, 48, 54
    frustration compared to, 48
    love and, 66
animal species
    bonding of, 234–37, 240–43, 240n, 241n, 243n, 253
    brain sizes of, 242–43
    care of offspring and, 32–33, 236
    consensual telepathy and, 273–74
    groups of, 243
    love associated with, 67, 234–35
    love of things and, 252–54
    motivational system bonding, 236
    responsive and, 71
    self-love and, 236–37, 277
    territory of, 240–41
anthropomorphic machines, 262
anthropomorphic objects
    consensual telepathy and, 272–73
    cookies as, 76, 76
    as honorary persons, 50
    human qualities assigned to, 43–48, 45, 47, 94, 134, 257, 259
    idols as, 255
    love of things and, 94
    products speaking to users and, 52–53, 272
    religious figurines as, 255

anthropomorphic objects (*continued*)
   trust and, 46–47, 50
   two-way relationship with, 70
anthropomorphic thinking
   anger at objects and, 48, 54
   hoarding and, 60–63
   human-looking objects having human
     qualities and, 43–48, *45*, 47, 85, 270
   loneliness and, 54–55
   love linked to, 48–51, 49n
   love of things and, 41–42, 70, 94
   marketing and, 51–54
   naming objects and, 50
   pets and, 55, 57–60
   problem solving and, 55–57
   situations leading to, 54–57
   talking to objects and, 48
   things disguising themselves as
     people, 42, 64, 251
anthropomorphism
   conversation generators and, 269–72
   as relationship warmer, 38, 39, 64, 94,
     106–7, 134, 248, 251, 252, 261
apes, 243, 243n
Apple
   brand love and, 23–24, 32, 80, 104,
     248, 261, 263
   Siri as voice of devices, 53
   tech sector and, 202n
appliances, ability to speak to users, 52
architecture, 21
Argo, Jennifer, 124
Argument, The (podcast), 213
Ariely, Dan, 128
aristocrats, 209–10, 219–20
Aristotle, 110, 244
Arnould, Eric, 256
Aron, Arthur, 66, 110, 114–15, 117–18,
   129, 132, 244
Aron, Elaine, 110, 114–15, 117, 129, 132
Arsena, Ashley Rae, 81
arts, love of, 9, 18, 187–88, 187n
Asian cultures, 140
asparagus rule, 210–11
atomic love, 149, 151–54
Atomic Love (film), 149n
atomic reactions, 151

attachment styles, 29–30
attachment theory, 29–30
attractiveness, 83
Austin-Healey Sprite, 44, *45*
authentic self
   collectivist culture and, 140
   family and, 139–40
   intrinsic rewards and, 147–49, 157
   intuitive fit and, 143–44
   project of the self and, 141–43, 157
   Romanticism and, 138, 138n, 139, 140

baby names, 170
Bach, Johann Sebastian, 168
Bagozzi, Rick, 10, 26, 119
ballroom dance clubs, 100
bargain hunting, love of, 17, 18–19
Bartneck, Christoph, 42
baseball, 185–86
Bates, Kathy, 198
Batra, Rajeev, 10, 26, 119
Beats, 268
Belk, Russell, 23, 88–89, 128, 247, 264
Bellagio resort and casino, Las Vegas, 82
Berger, Jonah, 99
Beverland, Michael, 60, 104
bird species, 235
bird-watching societies, 100
Birkin bag, 225–26
Blackwell, Ron, 246–47
Bloch, Peter, 128
BMW, 166
bohemians, 211, 211n, 215–16
bonding
   of animal species, 234–37, 240–43,
     240n, 241n, 243n, 253
   boundary-breaking experiences and,
     132–33
   brand love and, 133
   long-term love and, 238–40
   pair-bonding, 235, 236, 238, 240–41
books
   of books, 121–22
   boundary-breaking experiences and,
     132
   challenges of, 206
   comic books, 14

as emotional comfort, 72
love of reading, 8
self-concept and, 129
stimulation of, 177
tastes in, 103, 117
boredom zone, *171, 172, 188,* 189
Bottega Veneta, 227
boundary-breaking experiences, 132–34,
276
Bourdieu, Pierre, 197, 212–13, 303n5
bourgeoisie, 210, 210n
Brady, Tom, 19
brain. *See also* anthropomorphic thinking
age-related changes in, 167
default modes of thought and, 34–35,
38–39, 42, 43
electrode surgically implanted in,
265–66, 268–69, 273–74
evolution of, 54, 230, 242–43, 246, 255
fusiform face area, 246–47, 308n25
group identity markers and, 103, 104
heuristics and, 84
humanlike things treated as human,
48–49
left inferior prefrontal cortex, 248
limbic system of, 191
love of people and, 34–35, 38–39, 42,
43, 44, 85
love of self, 39
love of things and, 36–37, 38, 39, 42,
48, 94
medial prefrontal cortex, 36, 44, 248
mental model of physical world and,
176–77
metaphorical thinking and, 84
negative stereotypes and, 36
neocortex, 235, 242–43, 246
physical changes in, 231
safety and, 113
self-concept and, 111–12, *112,* 113,
118, 119
self-love and, 233–34
size of human brain, 242–43, 245
social brain thesis, 34, 38, 246–47, 248
sorting mechanisms of, 251
tools for thinking about people and,
54, 56, 57, 84, 248–52

tools for thinking about things and,
56, 68, 94, 248–52
unconscious predictions of, 176–77
brain-computer interfaces
consensual telepathy and, 273
consumer demand for, 267–69
products embedded with, 261, 262
self-integration and, 262–69
types of, 264–66
uses of, 266–67
brand communities, 101
Branden, Nathaniel, 147
brand love
anthropomorphic thinking and, 52–54
aspirational purchases and, 74–75
attachment styles and, 30
bonding experiences and, 133
brain regions activated by, 37
carryover effects and, 81
designer fashion leagues, 221–27, 223
disloyalty and, 80, 92–93
exciting/fling brands and, 82, 83, 238
as group identity marker, 101–3,
101n, 102n, 104
identity expressed through, 24, 25
lover's quarrel and, 82–83
luxury brands and, 23–25, 74, 154,
156–57, 202–3, 216–17, 218, 219,
221–27, 223
oxytocin and, 36
pain-reducing effect of, 37
personality of brands, 81–83, 91–92,
248, 259
person-thing-person connections and,
88, 91–93
public image and, 25
relationships and, 69, 73
religious aspects of, 23
responsiveness and, 71
scientific data on, 4
sincere/keeper brands and, 82, 238
Brasel, S. Adam, 82
bridge (card game), 174, 179
Brin, Sergey, 9
Broadbent, Sarah, 66, 133
Bronies, as brand community, 101–3,
101n, 102n

Bronies (documentary), 101, 102
brood parasites, 252–53
Brown, Lucy, 66
Bunam, Rabbi Simcha, 21–22

CafeDirect (fictional company), 46–47, 47
capital. *See also* cultural capital; economic capital
definition of, 197
capitalism, 214
card games, 174, 179
care packages, 19
Carey, Gregory, 168
carryover effects
relationships and, 81–84
salivation and, 230
cars
ability to speak to users, 52
anthropomorphic thinking and, 62
conversation generators and, 271–72
development of, 268–69
economic capital and, 216
feelings generated by, 98
human personality traits assigned to, 44–46, 45, 50, 50
identities and, 136
love of, 9, 18, 41
lowrider custom-car culture, 208–9
people talking to, 48, 55, 57
social status and, 196
as symbol of aspirations, 123
tastes in, 165–66, 166n
willingness to invest resources and, 31, 128
Castaño, Raquel, 37
Cast Away (film), 54–55
Cathy (comic strip), 149–50
cell phones
ability to speak to users, 52
integration into self and, 263
love of, 9, 18, 20, 32, 261
as people connectors, 39, 106
challenges
complexity of, 180–81, 190, 206, 207
in fun and flow zone, *171*, 172, 173, 174, 175, *188*

intensity of stimulation and, 180, 182–85, 206, 207
repetition and, 188–89
role in flow theory, 175–80
specialized knowledge required for, 180, 185–88, 206
subtlety of, 180, 181–82, 190, 206, 207
Chan, Cindy, 99
Chandler, Jesse, 259
Chanel, as luxury brand, 24, 152, 218, 225
Cheese Science Toolkit, 181
Chen, Rocky Peng, 51
children
altruistic caregiving from parents, 79
anthropomorphic thinking of, 54
emotional comfort of love objects and, 72
love associated with, 67
person-dog relationship type 1 and, 60
tastes of, 160–64, 166, 169
Chinese Confucian values, 153
Chopin, Frédéric, 189
Christakis, Nicholas, 17
Churchill, Winston, 9
Clancy, Stephanie, 245
Clode, Dani, 266
clothing
avant-garde fashions, 206, 207
brand love and, 25
designer fashion leagues and, 221–27, 223
fashion cycles, 170, 170n, 188, 189
good taste and, 205–6
love of, 9, 18
ready-to-wear collections and, 206–7
Coach, 222
Coaston, Jane, 213
cognitions, 26
collective resources, allocation of, 246
collectivist cultures, 140, 153, 204–5
collectors, love of things, 9
color, pleasant emotional memories and, 191–92
computers
anthropomorphic thinking and, 62
people's anger directed toward, 54

people talking to, 48, 55–56, 57
    safety of objects and, 72–73
consciousness, 111–12, 114, 115–16, 269
consensual telepathy
    people connectors and, 272–76
    products embedded with, 261, 262
conservatism
    local cultural capital and, 213–14
    values of, 209–12
Consiglio, Irene, 80
consumer behavior
    identities and, 155–56
    Industrial Revolution and, 136–37
conversation generators
    anthropomorphism and, 269–72
    products embedded with, 261, 262
Coon, Hilary, 168
Cottle, Michelle, 213
country, love of, 20
COVID-19 pandemic, 35
Crazy Rich Asians (film), 146–47
creativity
    cultural capital and, 205, 210, 211,
        211n, 226
    love objects and, 127–28, 255
cricket, 197
Csikszentmihalyi, Mihaly, 171, 171n
cultural capital
    counterfeit goods and, 202–3, 221, 227
    creativity and, 205, 210, 211, 211n,
        226
    cultural omnivores and, 211–12
    debates on, 207–17
    designer fashion leagues and, 221–27,
        223
    economic capital and, 201–2, 208,
        213, 214–17, 215, 227
    expertise and, 206–7
    general versus local cultural capital,
        208–9, 211, 213–14, 227
    hierarchies of, 212–14
    high cultural capital, 211–12, 215,
        216, 217, 219, 219n, 220, 226
    levels of, 195, 197, 198–99, 203–7
    love of things and, 197, 203, 205–7,
        227–28
    low cultural capital, 215, 218

shift from conservative to liberal
        values, 209–12, 213
    social status and, 197, 198–99, 201–2,
        209, 212–13, 214, 215, 220
    wealth and, 201–2, 226–27
Cultural Capital Quiz, 199–200
cultural creatives, 214–17, 215
cultural diversity, 211
culture wars, 207–8, 209
current identity, 27, 31–32
cuteness expression, 67
Cutright, Keisha, 24

Dahl, Darren, 74, 124
Dalton, Pamela, 191
dating behavior, 14, 83
dating services, 3–4
De Angelis, Barbara, 20
decluttering, 78, 98–99
Deezer, 167
dementia, 169
desired identity, 27, 31–32
diabetes, 185
DiCaprio, Leonardo, 198
dogs
    anthropomorphic thinking and, 58–59
    love of things and, 254
    person-dog relationship type 1, 59–60
    person-dog relationship type 2, 59–60
    responsiveness of, 71
    robotic dogs, 270
    talking to, 58–59
Dollinger, Stephen, 245
Downton Abbey (television show), 209
drug addiction, 182, 183
Dunbar, Robin, 243
Dunn, Lea, 133
Durante, Kristina, 81

economic capital
    counterfeit goods and, 202–3, 221, 227
    cultural capital and, 201–2, 208, 213,
        214–17, 215, 227
    debates on, 214–17
    designer fashion leagues and, 221–27,
        223
    display of, 217–20, 219n

economic capital (*continued*)
  high economic capital, *215*, 216, 217, 219–20, 226–27
  levels of, 195, 197–98, 203–7
  love of things and, 197, 203–5, 227–28
  low economic capital, 215, *215*, 216, 217–18
  social status and, 197–98, 201, 213, 214, 215, 216
education
  cultural capital and, 201–2, 207–8, 211
  economic capital and, 201–2, 213, 214
EEG (electroencephalograph) technology, 265
egalitarianism, 211
egoism, 243
Eilish, Billie, 169
elitism, 211
Ellie, as virtual interviewer, 270
Ellis, Albert, 32
emotional experiences
  length of, 66
  love as, 66–68, 69
  of love objects, 68–69, 132–34, 170–71
  pleasant emotional memories, 190–92
  self-referential emotions test, 113–14
  types of, 27–28
Emotiv, 265
empathy, 44, 67, 68–69
employer-employee relationship, 78
empowerment, 21–22
entertainment
  complexity of challenges in, 180–81, 190, 206, 207
  cultural capital and, 210, 211
  intensity of stimulation and, 180, 182–85, 206, 207
  repetition and, 188–92
  role of challenges in, 175–80
  specialized knowledge required for, 180, 185–88, 206, 207
  subtlety of challenges in, 180, 181–82, 190, 206, 207
Erdem, Tülin, 24
Erikson, Erik, 141–42

euchre, 174, 179
Europe, Industrial Revolution in, 138
European common cuckoo, 253
evolution
  anthropomorphic metaphor for, 37, 37n
  behavior as evolutionarily optimal, 257–59
  bonding within animal species and, 235–37
  of brain, 54, 230, 242–43, 246, 255
  of friendship, 242, 244–46, *244*, 257–58
  of love, 230, 235–37, 244–45
  of love of people, 32–33, 34, 37, 37n, 38, 257
  of love of things, 34, 37, 37n, 38, 38n, 257–59, 261
  sexual attraction and, 231, 232
evolutionary biologists, 237n
exercise, 72, 129
exhilaration, 27–28
expectations, 14, 32, 78, 79, 177, 272
expertise, levels of, 120–21, 206–7
extreme couponing, 18–19
Extreme Couponing (television show), 19
extrinsic rewards, 145, 146, 147, 148–49, 157

Facebook, 271–72
family
  authentic self and, 139–40
  love and, 243, 244, 257, 258–59
  person-thing-person connections and, 88–92, 95, 98
  photos as people connectors, 39, 97, 98
  sense of self and, 234–41, 237
Farquhar, Sarah, 125
fashion cycles, 170, 170n, 188, 189
fear, 132–33
fearful attachment style, 30
Feinberg, Fred, 248
feminism, 140, 152
fight-or-flight mode, 58
fish, reproduction of, 231, 231n
Fisher, Helen, 66, 238, 305–6n2

Fiske, Susan, 36, 248
Fitzsimons, Gavan, 24
Fletcher, Garth, 275
flow theory
    activities and, 173–74, 175, 176, 177,
        180
    boredom zone, *171*, 172, 189
    diagram of, *171*, 172
    flow in unexpected places, 175–80
    flow state as peak experience, 171
    frustration zone, 89, *171*, 172, 179,
        188, 189
    fun and flow zone, *171*, 172–73
    repeated experience and, 188, *188*, 189
    repetition and, 188–92
    role of challenges in, 175–80
    role of goals and feedback in, 175
Flynn effect, 168
fMRI machines, 48, 265, 265n
food
    carryover effects and, 230
    connecting power of, 105
    cultural influence on, 159–60
    desserts and, 238–39
    emotional comfort and, 72
    identities and, 126
    intensity of stimulation and, 182, 183,
        184–85, 206
    love of, 238–39
    pleasant emotional memories and,
        190
    subtlety of, 181–82, 206
    tastes formed in childhood, 162–64,
        166, 169
Force Trainer II: Hologram Experience,
    265
Ford, Tom, 155–56
Ford GT supercar, 74
Foster, Scott, 162
Fournier, Susan, 69, 82
Fox News, 213
France, 210, 210n
French Revolution, 210
friendship
    consensual telepathy and, 275–76
    evolution of, 242, 244–46, *244*,
        257–58

gift exchange and, 256
    moral obligations and, 78
    mutual-aid pact created by, 258
    person-dog relationship type 2 and, 60
    person-thing-person connections and,
        89, 92–94, 96–97
Fromm, Erich, 110, 114, 127, 276
frustration
    anger compared to, 48
    zone of, 89, *171*, 172, 179, *188*, 189
fun and flow zone, *171*, 172–73, *188*
Fürst, Andreas, 36
fusiform face area, 246–47, 308n25

Gal, David, 229
game design, 173
gardening, 8, 67, 114
Gates, Bill, 263
gender norms, 102
gender roles, 138, 139, 149–52
generosity, 247
genetics
    intensity of stimulation and, 182
    love of people and, 257, 258, 259
gifts
    exchange of, 247, 255–56
    as people connectors, 39, 98
    as relationship markers, 98, 99
Gillath, Omri, 193
God
    grace and, 15
    love of, 8–9, 18, 20, 26
Goldstein, Pavel, 276
golf, 174
Gonzalez, Richard, 81
Goode, Miranda, 93
Google, 143, 202n
grace, as religious concept, 15
groups
    effective teams and, 243–44
    group identity markers, 99–105, 101n,
        102n, 242
    lifestyle groups, 194–96, 217–21, 227
    person-thing-person connections and,
        88
    sense of self and, 241–52, *244*
    social intelligence and, 245–46

Gucci, as luxury brand, 23, 24, 222, 224, 225
Gucci and Balenciaga co-branded Marmont bag, 223
Gucci Tifosa bag, 223, 226
Guisewite, Cathy, 149–50, 151
Gutchess, Angela, 248

habituation, 183–85
Haire, Mason, 135
Hallmark, 81–82
Hanks, Tom, 9, 54–55
Harley-Davidson riders' groups, 100
Harlow, Harry, 254
Harris, Lasana, 36
Hatfield, Elaine, 78
Haute Couture League, 223, 226–27
HBO, 100
head coverings, as group identity markers, 99
hearts (card game), 174
heirloom jewelry, as relationship markers, 98
Hello Barbie, 270–71
Hermès Verrou ostrich leather bag, 223
heuristics, 83–84
Hidden Brain (podcast), 55
high cultural capital, 211–12, 215, 216, 217, 219, 219n, 220, 226
high economic capital, 215, 216, 217
hippies, 193, 215
hipsters, 215
Hirschberg, Lynn, 155
Hitler, Adolf, 124
hoarding, 60–63
hobbies, 9, 65
Hoegg, JoAndrea, 133
Homer, 208
homes
    caring for one's home, 128
    intuitive fit and, 29
    love of, 9, 18, 41
    as people connectors, 39
Homo erectus, 255
Horyn, Cathy, 206
Hudson, Simon, 52–53

human-looking objects, 43–48, 45, 47, 85, 270
humility, 21–22
Hur, Julia, 76

iCat talking robot, 42–43, 42
identities
    in adolescence, 165
    advertising and, 154–57, 155n
    authentic self and, 138–42, 138n
    collectivism and, 204
    conflict between identities, 149–54
    cultural versus economic capital and, 216
    defining of identity, 142, 204
    drawing distinctions and, 103–4
    food and, 126
    group identity markers, 99–105, 242
    individualism and, 157, 204
    insights into, 6
    love as fusion of, 110–11, 114, 115, 116, 117, 119
    love of things creating, 5, 24, 109–10, 143–44
    love of things expanding, 27, 31–32, 36, 38, 39, 106, 129, 130, 251, 269, 276
    love of things expressing, 18, 24–25, 116, 136, 216, 227
    possessions and, 130–31
    Romantic Era and, 138–39, 140
    self-concept and, 111–14, 119–21, 245
    self-descriptive words representing, 129–30
identity crisis, 142
IKEA effect, 128
individualism
    cultural capital and, 203
    economic capital and, 204
    project of the self and, 141
    Romantic Era and, 137–38, 140
    wealth in cultures and, 137–38, 140, 157, 204
Industrial Revolution, 136–38, 157, 220
infancy and childhood, tastes in, 160–64, 166, 169
in-groups, 104

insula, 37
internet, 53
interpersonal love. *See also* groups;
    romantic partnerships
    studies of, 4, 120
interpersonal primacy, 240
intrinsic rewards, 144–49, 157
intuitive fit, 27, 29, 143–47
Inuit, 137
investment
    altruism versus, 79–80
    in love objects, 127–28, 255
IQ, 168

Jack Armstrong factor, 14
James, William, 130
jealousy, 28
Jobs, Steve, 23
Johnson, Allison, 30, 69
Joubert, Joseph, 15
Journal of Neuroscience, 246
Just, Marcel, 265

Kaplan, Mordecai, 101
Keller, Helen, 114
Kervyn, Nicolas, 248
Khamitov, Mansur, 93
Kim, Sara, 43, 51
Kingston, Karen, 99
Kitayama, Shinobu, 140
knitting circles, 100
Kondo, Marie, 63, 227
Konrath, Sara, 118
Kotler, Philip, 3–4
K reproductive strategy, 235–36
Krumhuber, Eva, 93
Kula gift exchange system, 255–56

Landon, Adam, 106
Landwehr, Jan, 45
lang, k.d., 191
Larson, Gary, 59
Lastovicka, John, 31
Lauren, Ralph, 25–26
Law & Order (television show), 178–79,
    180

Lee, David Seungjae, 81
legal ownership, 130–31
leisure time, 197–98
LGBTQ+ movement, 140, 213
liberalism, values of, 193, 209–12, 213
libertarians, 214
lifestyle groups
    display of cultural capital and, 217,
        218, 227
    display of economic capital and,
        217–21, 227
    principles giving rise to, 194–96
Lifshitz, Ralph, 25–26
like/love distinction, 18–20
Linden, David, 236
local cultural capital, 208–9, 211,
    213–14, 227
Logo League, 221–22, 223, 224, 226, 227
loneliness, 54–55, 63, 73, 98, 106
long-term goals, 58
Louis Vuitton, 222
love
    anthropomorphic thinking linked to,
        48–51, 49n
    atomic love, 149, 151–54
    carryover effects and, 230
    consensual telepathy and, 273
    decision to love, 15
    as deep, profound experience, 18–20
    defining of, 231–32
    as emotional experience, 66–68, 69
    evolution of, 230, 235–37, 244–45
    family and, 243, 244, 257, 258–59
    as fusion of identities, 110–11, 114,
        115, 116, 117, 119
    as goal-directed state, 66
    long-term love system, 238–40
    meanings of, 67, 78, 84
    persistence and, 32
    psychology of, 4–5, 7
    Romantic Era and, 138–39, 231
    sacredness of, 20–22
    theories about, 110
    as type of relationship, 68–69
    types of, 12
    unconditional love, 15–16, 17

love at first sight, 143–44
love objects
 boundary-breaking experiences and,
  132–34, 276
 brain regions and, 36–37
 buying and owning of, 130–31
 closeness and, 77–78
 creation and investment in, 127–28,
  255
 emotional comfort of, 72
 emotional experiences of, 68–69,
  132–34, 170–71
 intrinsic rewards and, 144–47
 intuitive control of, 126–27
 life stories including, 121–23
 logistical support and, 105–6
 as people connectors, 90–91, 94–97,
  242
 physical contact with, 124–26
 projective questions about, 116–17
 safety of, 72–76
 self-integration of, 27, 31–32, 267,
  269, 276
 sense of self and, 34, 68, 110, 115
 things as, 6
 thinking and learning about, 119–21
 tragic if lost, 34
 two-way relationship with, 70, 72
 virtues of, 13–14
love of money, 93, 257
love of people
 brain and, 34–35, 38, 42, 43, 44, 85
 emotional attachment and, 29–30
 frequent thoughts and, 120
 genetics and, 257, 258, 259
 incorporated into self-concept, 114,
  115, 117, 118–19, 134
 jealousy associated with, 28
 like/love distinction and, 18
 long-term relationships and, 33
 love of things compared to, 5, 12,
  15–16, 17, 20, 28–30, 31, 33, 34, 84,
  85, 240–42, 257–59, 270, 276
 objectification of people as opposite
  of, 35–36, 93
 as one-way relationship, 70, 84
 oxytocin and, 36

psychological need and, 17
 as two-way relationship, 69, 84
 unconditional love for, 15–16
love of things. *See also* brand love; love
  objects
 animal species and, 252–54
 anthropomorphic thinking and,
  41–42, 70, 94
 approach to romantic partnerships
  compared to, 13–14
 attachment styles and, 30
 attributes of, 26
 beautiful things, 232–33
 carryover effects and, 230
 character revealed by, 193–94
 chemistry and, 144–47
 components of, 26, 27
 cultural capital and, 197, 203, 205–7,
  227–28
 development of, 231, 252–57
 economic capital and, 197, 203–5,
  227–28
 emotional attachment and, 27, 29–30,
  63–64, 94, 170–71
 emotional comfort and, 72
 evolution of, 34, 37, 37n, 38, 38n,
  257–59, 261
 excellence and, 13–14, 15, 17, 26, 27
 expertise about things, 120–21, 206
 flow state and, 171, 192
 frequent thoughts about things, 27,
  32, 119–20, *120*
 as group identity markers, 100
 humblebrags and, 17
 identities created by, 5, 24, 109–10,
  143–44
 identities expressed by, 18, 24–25,
  116, 136, 216, 227
 incorporated into self-concept,
  109–10, 114, 115, 118–19, 134, 234,
  253
 in-groups and out-groups
  distinguished in, 104–5
 interpersonal primacy and, 240
 intrinsic rewards and, 144–47
 intuitive fit with, 27, 29, 143–47
 judgments of quality and, 16, 17

lack of unconditional love for things, 15–16, 17
like/love distinction and, 18–20
local cultural capital and, 208–9
long-term relationship and, 27, 32–33
love as emotion and, 68
love of flawed things connected to love of people, 16
love of people compared to, 5, 12, 15–16, 17, 20, 28–30, 31, 33, 34, 84, 85, 240–42, 257–59, 270, 276
low tolerance for imperfection in, 16–17
lust and, 231–33
meaningful experiences and, 18, 19–20, 32, 161–62
as metaphorical love, 12, 20
moral obligation and, 78–80
as one-way relationship, 69, 84
passionate involvement and, 27, 30–31
passion in, 8, 121, 143, 149–54, 232–33
past involvement and, 27, 31, 32
positive emotional connection and, 26–30, 27
practical benefits and, 15, 37, 256, 258–59
psychological visibility and, 147–49
psychology of, 4–5, 7
reduction in objectification and, 35–36
responsiveness and, 71–72
self-integration of love object, 27, 31–32, 267, 269, 276
sense of self and, 109, 252–57, 253, 277
shaped by relationships to objects and activities, 12
special possessions, 89
tragic if lost and, 27, 33–34
as two-way relationship, 70–71
willingness to invest resources and, 27, 31
Love of Things Quiz, 10–11, 10–11, 26, 161–62
love thermometer scoring guide, 11
love welling, 67–68

low cultural capital, 215, 218
low economic capital, 215, 215, 216, 217–18
Lucite handbags, 152
lust
  love of food and, 238–39
  love of things and, 231–33
Luxe League, 223, 224–26, 227

McGill, Ann L., 43, 46, 51–52
Madden, Thomas, 12
Made for Love (HBO), 275
Madonna factor, 14
mainstream elites, 214–17, 215
Makin, Tamar, 266
Malinowski, Bronislaw, 255
Malone, Chris, 248
mammals, love experienced by, 235, 236
Manolo Blahnik, 224
manipulation, 43
marketing. See also advertising
  anthropomorphic descriptions for products, 63
  anthropomorphic thinking and, 51–54
  dating compared to, 3–4
  exclusivity of brands and products and, 74
  lifestyle groups and, 194–95
  money spent by advertisers, 8
  product family versus product line, 52, 52
  scarcity of brands and products and, 73–74
  of shampoos, 125–26
Markus, Hazel, 140
marriages, 14, 71, 138–39, 141
Maslow, Abraham, 110
Masterpiece (television show), 212
materialism, 93, 215, 219
Maxian, Wendy, 26
meaningful experiences
  flow and, 176
  intrinsic rewards and, 146
  life stories including love objects and, 121–23

meaningful experiences (*continued*)
love of things and, 18, 19–20, 32,
161–62
people connectors and, 95, 272
purpose in life and, 18, 57
specialized knowledge and, 187
mechanical exoskeletons, 263, 266
medieval cathedrals, 21
mementos, as relationship markers, 89,
98
memory bias, 118–19
men
importance of responsiveness to, 71
Jack Armstrong factor and, 14
on sexism, 212
Mennella, Julie, 164
mental capacity, brain-computer
interface and, 266–67
metaphorical thinking, 84
Mexican American community, 208–9
Meyers, Anne Akiko, 87–88
Michael Kors, 222
Michelangelo, 89
middle age and later, tastes in, 166–70
mirror neurons, 44
Mitchell, Tom, 265
Mochon, Daniel, 128
Modrak, Rebekah, 25
money. *See also* wealth
as extrinsic reward, 146
love of, 93, 257
as representation of power and status,
229–30
Monga, Alokparna, 71
monkeys, 243, 243n, 254, 254n
Moon, Youngme, 70
Morales, Andrea, 124
morality
altruism versus investment, 79–80
anthropomorphic thinking and,
62–63
anthropomorphized products and,
53
helping other people and, 109
monogamy and, 80
relationships and, 78–80
Mosaic, 194–95

motorcycles, 62
Mozart, Wolfgang Amadeus, 168
Mr. Clean, 51
Murstein, Bernard, 14, 239
music
challenges and, 207
emotional comfort and, 72
exhilarating emotional experiences
and, 27
flow state and, 175, 176, 179–80,
188–89
global music industry, 169
group identity markers and, 103, 105
intensity of stimulation and, 183
love of, 8, 22
nature and nurture playing role in
musical ability, 167–69
as people connectors, 39
pleasant emotional memories and, 191
repetition and, 188–89
taste in, 164, 165, 167, 169
musical instruments, 71–72, 79, 87–88
Musk, Elon, 274–75
My Little Pony (television show), 101–3,
102n

National Firewood Night (television
show), 9–10
National Institutes of Health, 57–58
nature
emotional comfort and, 72
love of, 8–9, 18, 20–21, 22, 26, 255
person-thing-person connections and,
106
protection of natural areas, 79
Ndumele, Chiadi E., 185
negative stereotypes, 36
Nemeroff, Carol, 124
Nescafé, 135–36
Nestlé, 135–36
Neuralink, 274–75
New Guinea, 255–56
New York Times, 24, 206, 213
New York Times Book Review, 121
Nicolelis, Miguel, 273
nonprofit organizations, 71
Norton, Michael, 128

Nuñez, Sandra, 37
Nussbaum, Martha, 35

Obama, Barack, 143
objectification of people, as opposite of
   love of people, 35–36, 93
Obstfeld, David, 183–84
Onassis, Jacqueline Kennedy, 124
online social interactions, 271
Oprah Winfrey Show, The (television
   show), 4
orbicularis oculi muscles, response of,
   26–27
out-groups, 104
Oxfam resale shop, Manchester,
   England, 124–25
Oxfam Shelflife project, 125
oxytocin, 36, 67, 170

pain research, 37
pair-bonding, 235, 236, 238, 240–41
Palmer, Stephen, 191–92
Papez, Maja Golf, 104
parenting, 242–43
Paris, Meg, 129
Park, C. Whan, 118
Parkinson, Carolyn, 275
passion, in love of things, 8, 121, 143,
   149–54, 232–33
passionate involvement, 27, 30–31
Passman, Richard, 269
Patek Philippe, 98
Paul, Pamela, 121–22
people connectors
   consensual telepathy and, 272–76
   group identity markers and, 99–105
   logistical support and, 105–6
   love objects as, 90–91, 94–97, 242
   person-thing-person connections and,
      89, 134, 272
   relationship markers and, 97–99
   as relationship warmers, 39, 94,
      106–7, 134, 248, 251, 252, 261
   types of, 97–106
people of color, 212, 213
perfume, advertising for, 154–55, 155n
personal growth, 129–30

person-thing-person connections
   Russell Belk on, 88–89, 247
   brand love and, 88, 91–93
   family and, 88–92, 95, 98
   friendship and, 89, 92–94, 96–97
   Kula gift exchange system and, 256
   love objects and, 89, 94, 257
   nature and, 106
   people connectors and, 89, 134, 272
   relationship markers and, 98
   relationships and, 88–93, 97, 98, 106,
      134, 187, 247, 251
   sense of self and, 251
   sports and, 187
person-thing-person effect, 92, 93
pets. *See also* dogs
   anthropomorphic thinking and, 55,
      57–60
   love of, 8–9, 18, 57
   relationship with, 73, 234
   sense of purpose and, 57–58
Pew Research Center, 22
Philip, Prince, 221
photos of friends and family, as people
   connectors, 39, 97, 98
Pilati, Stefano, 206
Pinto, Juliano, 263, 266
Pirates of the Caribbean (film), 142
Plutarch, 257
Polk, Thad, 248
popular culture, 212
Porsche, 82, 166
positive feelings, indicators of, 27, 27
post-traumatic stress disorder, 270
Powell, Elyse, 184
Prada, 222, 224
Praxiteles, 89
pride, 21, 104, 113–14, 117, 205
problem solving, 55–57
product quality, 49n
Progressive Insurance, 41
progressive political views, 211, 213,
   215–16
projective questions, 31–32, 116–17
project of the self, 141–43, 157
prosthetics, 264, 266–67, 267
psychological ownership, 130–31

psychological visibility, 147–49
psychologists, 237n
purpose in life, 18, 57
Pyer Moss, 227

Racine, Jean, 82n
racism, 212
Ralph Lauren brand, 25–26
Ranganathan, Aruna, 79
Rappaport, Amelia, 163
Rapson, Richard, 78
Rauschnabel, Philipp, 49, 97, 106
reality television, 212
recreational objects, 98
Reddit, 35, 41, 124, 262
Reik, Theodor, 110
Reimann, Martin, 36–37
Relationship Closeness Inventory,
    77–78
relationship markers, 97–99
relationships
    abusive relationships, 75
    brand love and, 69, 73
    carryover effects and, 81–84
    closeness and, 5, 76–78, 84
    consensual telepathy and, 275–76
    defining of, 65
    emotional comfort and, 71, 72
    give and take of, 272
    love as type of relationship, 68–69
    marriages, 14, 71, 138–39, 141
    moral obligation and, 78–80
    one-way relationships, 69, 70
    oxytocin associated with, 67
    people who have passed away and,
        73n
    person-thing-person connections and,
        88–93, 97, 98, 106, 134, 187, 247,
        251
    pets and, 73, 234
    responsiveness and, 71–72
    safety of objects and, 72–76
    two-way relationships, 69, 70–71
relationships with things, meaning of, 65
relationship warmers
    anthropomorphism as, 38, 39, 64, 94,
        106–7, 134, 248, 251, 252, 261

people connectors as, 39, 94, 106–7,
    134, 248, 251, 252, 261
sense of self as, 39, 134, 248, 251, 252,
    261, 262–69
relaxation, as emotional experience, 27,
    28
religion
    attitudes toward, 22
    on consciousness as part of soul, 111
    food restrictions of, 126
    as group identity marker, 100–101
    identity expressed through, 24–25,
        113, 138
rental car companies, 62
reproduction
    of fish, 231, 231n
    K reproductive strategy, 235–36
    r reproductive strategy, 231, 233–34,
        235, 236
reputations, 5
responsiveness, 71–72
Richins, Marsha, 98
robots, 55, 62, 270
Rodas, Maria, 173
Rodin, Auguste, 89
Rolex watches, 153–54
Romantic Era, 138–39, 138n, 140, 231
romantic partnerships
    consensual telepathy and, 276
    exaggerating good qualities of partner
        and, 13–14
    interpersonal assets and liabilities in,
        239–40
    marriages, 14, 71, 138–39, 141
    social skills and, 245
Roma peoples, 211n
Rosen, David, 165, 301n2
Ross, Michael, 118
Rousseau, Jean-Jacques, 233
Rozin, Paul, 124
r reproductive strategy, 231, 233–34, 235,
    236

salivation, 229–30
Sandler, Adam, 132
Sapolsky, Robert, 167
Sartre, Jean-Paul, 121

scary movies, as bonding experience,
132–33
scent, emotional memories triggered by,
191
Schloss, Karen, 192
Schwarz, Norbert, 259
secure attachment style, 30
Seinfeld, Jerry, 24, 263
self. *See also* authentic self; sense of self
consciousness equated with, 111, 114,
269
fuzzy boundary around, 113, 276
project of the self, 141–43, 157
self-actualization, 141
self-concept
boundary-breaking experiences and,
132–34
brain and, 111–12, *112*, 113, 118, 119
buying and owning love objects and,
130–31
consciousness and, 111–12, 114,
115–16
continuum of, 112–13, *112*
control of love objects and, 126–27
creation and investment in love
objects and, 127–28
expansion of, 236–37, *237*, 244, *244*,
252–53, *253*
falling in love affecting, 129–30
identities and, 111–14, 119–21, 245
interdependent self-concept, 140
life stories including love objects,
121–23
love objects changing you and,
129–30
love of people incorporated into, 114,
115, 117, 118–19, 134
love of things incorporated into,
109–10, 114, 115, 118–19, 134, 234,
253
physical contact with love objects
and, 124–26
self-love and, 231–34, 233
self-referential emotions test and,
113–14
thinking and learning about love
objects, 119–21

self-discovery, 148–49
self-esteem, 233
self-image, 111
self-integration
aspects of, 27, 31–32, 261, 267
brain-computer interfaces and,
262–69
love objects included in life stories
and, 121–23
mental processes and, 119–21
self-love, 231–34, 233, 236–38, 252,
277
self-referential emotions test, 113–14
sense of self. *See also* authentic
self; identities; self-concept;
self-integration
brain-computer interfaces and,
262–69
expansion of, 237, *237*, 244–45, *244*,
252, 253, 277
family and, 234–41, 237
friendship and, 242, 244–46, *244*,
257–58
groups and, 241–52, *244*
love objects and, 34, 68, 110, 115
love of people and, 109
love of things and, 109, 252–57, *253*,
277
objects of hoarders integrated into, 62
prosthetic self and, 264
as relationship warmer, 39, 134, 248,
251, 252, 261, 262–69
self-love and, 231–34, 233
service providers, 73
sexism, 212
sexual attraction, 231–33, 238–40
sexual desire, 69, 238
sexual relations, 121
Shachar, Ron, 24
Shakespeare, William, 208, 257
shampoos, marketing of, 125–26
Shankar, Maya, 142–43
Shimp, Terence, 12
shoes, character revealed by, 193–94
shopping, 28, 144
short-term goals, 58
Sirianni, Nancy, 31

skills
    factors in decrease of, 179
    in fun and flow zone, *171*, 172, 174,
        176, 188–89, *188*
    social skills, 245
slot machines, images of, *43*, *43*
Smith, Adam, 137
social brain thesis, 34, 38, 246–47, 248
social hierarchy, 46, 210
social intelligence, 245–46
social media
    number of followers, 74
    products speaking to users and, 52–53,
        271–72
social status
    aristocrats and, 209–11, 220
    bourgeoisie, 210, 210n
    collectivism and, 204–5
    cultural capital and, 197, 198–99,
        201–2, 209, 212–13, 214, 215, 220
    economic capital and, 197–98, 201,
        213, 214, 215, 216
    individualism and, 204
    status symbols and, 196, 217–21
Socrates, 110
Sony Walkman, 268
Sood, Sanjay, 233
space aliens, 162n
SpaceX, 275
spades (card game), 174
spirituality, 22–23
sports
    animal mascots used by sports teams,
        58
    boundary-breaking experiences and,
        133–34
    colors and, 192
    emotional devastation of big losses
        and, 28–29
    love for sports teams, 66
    love of, 9, 18
    specialized knowledge of, 185–86, 187
Sposito, Ambra, 263–64
Steiner, Peter, 271
Stephane Rolland, 227
stereos, 98
Sternberg, Robert, 15

Stilley, Karen, 91
stress, 58
Subaru BRZ, 44, *45*
subcultures, 102n, 194, 208, 211, 211n,
    214, 227
Subway bread, 185
successes and failures, memory bias in,
    118–19
sushi, 167
Swaminathan, Vanitha, 30, 83, 91

Target, 25
tasks, difficulty level of, 172, 173
tastes
    in adolescence through early
        adulthood, 164–66, 167, 169
    in books, 103, 117
    in cars, 165–66, 166n
    cultural influence on, 159–60
    development of, 169–70, 195
    "good taste," 205, 205n
    group identity markers and, 103–5
    in infancy and childhood, 160–64,
        166, 169
    influence of other people on, 196
    in middle age and later, 166–70
    wealth and, 204, 219, 220–21
technology. *See* brain-computer
    interfaces; consensual telepathy;
    conversation generators
tech sector, 202n
Teletubbies (television show), 164, 178,
    180
television
    emotional comfort and, 72
    reality television, 212
    skilled activity involved in, 178–79
    watching habits and, 272
Terence, 82n
Tesla, 275
testosterone, 238
things. *See also* love of things
    activities as, 5, 7, 8
    animals as, 5–6
    disguised as people, 42–43
    love objects as, 6
    role of, 5

Third Thumb, 266, 267
Thomson, Matthew, 30, 69, 93
thought, default modes of, 34–35
thought recognition device, 265
tiredness, 179
Titanic (film), 198
Tito, Josip Broz, 221
Toobin, Jeffrey, 121–22
tools, 242–43
Torelli, Carlos, 173
Touré-Tillery, Maferima, 46
Townsend, Claudia, 233
Toyoda, Akio, 271
Toyota Friend, 271
transgressive behavior, 102–3
traveling, 28
Trump, Donald, 219n
trust
    anthropomorphic objects and, 46–47,
        50
    oxytocin associated with, 67
Tumbat, Gülnur, 23
twin studies, 168

University of Utah, 185
Urban, Tim, 274–75

Valve Corporation, 173–74
Van Boven, Leaf, 99
Versailles, 219
video games
    anthropomorphized characters with
        hints in, 51
    flow zone and, 173–74, 175
    as relationship markers, 98
Vieuxtemps, Henri, 87–88
Vieuxtemps violin, 87–88
voles, 234–35, 240–41, 240n, 241n

Wallendorf, Melanie, 256
Walmart, 7
Wan, Jing, 51
Wang, Xijing, 93
war (card game), 174

Ward, Morgan, 74
Wason selection task 1, 249–51, 249
Wason selection task 2, 249–51, 250
Watanuki, Shinya, 119
waterskiing, 28
Waymo Firefly (Google), 50, 50
wealth
    cultural capital and, 201–2, 226–27
    generational wealth, 98
    individualism and, 137–38, 140, 157,
        204
    progressive bohemians on, 215–16
    tastes and, 204, 219, 220–21
Western culture, influences of, 140,
    141–42
Whelan, Jodie, 30, 73
whist (card game), 174
Wierzbicki, Krzysztof, 162
Wilson, Carl, 207
Wilson, E. O., 243
Wilson, Timothy D., 251
Windhager, Sonja, 44
wine
    cultural capital and, 203
    specialized knowledge of, 186–87, 203
Wired magazine, 101
women
    carryover effects and, 81
    importance of responsiveness to, 71
    Madonna factor and, 14
    person-dog relationships and, 60
    on sexism, 212
Wood, Jeremy, 73n
work, flow state in, 174–75
World Cup, 263
Wozniak, Steve, 9, 23

Ybarra, Oscar, 81
Yoon, Carolyn, 248
Young, Larry, 240–41
Young, Neil, 191
Yves Saint Laurent, 206

Zhang, Ke, 51

About the Author

Aaron Ahuvia, PhD, is the world's most widely published and cited academic expert on noninterpersonal love, including brand love. He is also a leading expert on the ways in which happiness is influenced by money and materialism. He has been ranked number 22 in the world for research impact in consumer behavior and ranked in the top 2 percent of scientists in all disciplines worldwide in an independent study conducted by Stanford University. Dr. Ahuvia studied philosophy at the University of Michigan before getting a PhD in marketing from Northwestern University's Kellogg School of Management. From there he became a professor at the University of Michigan's Ross School of Business and is now the Richard E. Czarnecki Endowed Collegiate Professor of Marketing at the University of Michigan–Dearborn College of Business. Dr. Ahuvia also holds an appointment as a professor at the University of Michigan's Penny W. Stamps School of Art & Design.

Dr. Ahuvia has more than one hundred academic publications and presentations to his credit. He researches, teaches, and consults for governments, nonprofit organizations, and corporations in China, Denmark, Oman, Finland, Poland, Morocco, France, Pakistan, Germany, India, Israel, Italy, Jordan, Kazakhstan, Portugal, Rwanda, Singapore, Slovakia, Switzerland, the Netherlands, and Yemen. He has presented research or consulted for Google, L'Oréal S.A., Samsung, Maybelline New York, Procter & Gamble, Audi, General Motors, Microsoft, Ford, Chrysler, GfK market research, and Herman Miller, among other firms.